Introduction to Autonomous Robots

Introduction to Autonomous Robots

Mechanisms, Sensors, Actuators, and Algorithms

Nikolaus Correll, Bradley Hayes, Christoffer Heckman, and Alessandro Roncone

The MIT Press
Cambridge, Massachusetts
London, England

The MIT Press would like to thank the anonymous peer reviewers who provided comments on drafts of this book. The generous work of academic experts is essential for establishing the authority and quality of our publications. We acknowledge with gratitude the contributions of these otherwise uncredited readers.

This book was set in Times by Westchester Publishing Services. Printed and bound in the United States of America.

Names: Correll, Nikolaus, author.
Title: Introduction to autonomous robots : mechanisms, sensors, actuators, and algorithms / Nikolaus Correll, Bradley Hayes, Christoffer Heckman, and Alessandro Roncone.
Description: Cambridge, Massachusetts : The MIT Press, [2022] | Includes bibliographical references and index.
Identifiers: LCCN 2022010036 (print) | LCCN 2022010037 (ebook) | ISBN 9780262047555 (hardcover) | ISBN 9780262372947 (epub) | ISBN 9780262372954 (pdf)
Subjects: LCSH: Autonomous robots.
Classification: LCC TJ211.495 .C67 2022 (print) | LCC TJ211.495 (ebook) | DDC 629.8/932–dc23/eng/20220801
LC record available at https://lccn.loc.gov/2022010036
LC ebook record available at https://lccn.loc.gov/2022010037

10 9 8 7 6 5 4 3 2 1

For
Arthur, Tatiana, Benedict, and Silvester,
David, Leonardo, and Lily,
future robot users.

Contents

Preface

This book provides an algorithmic perspective on autonomous robotics to students with a sophomore-level knowledge of linear algebra and probability theory. Robotics is an emerging field at the intersection of mechanical engineering, electrical engineering, and computer science. With computers becoming more powerful, making robots smart is a major focus of attention, and robotics research is the most challenging frontier. While there are many textbooks on the mechanics and dynamics of robots available to sophomore-level undergraduates, books that provide a broad algorithmic perspective are mostly limited to the graduate level. This book was not created to be "yet another textbook that is better than the others." Rather, it was written specifically to teach robotics to the third- and fourth-year undergraduates at the Department of Computer Science at the University of Colorado.

Although falling under the umbrella of "artificial intelligence," standard AI techniques are not sufficient to tackle problems that involve uncertainty, such as a robot's interaction in the real world. This book uses simple trigonometry to develop the kinematic equations of manipulators and mobile robots, then introduces path planning, sensing, and, lastly, uncertainty. The robot localization problem is introduced by formally defining error propagation, which leads to Markov localization, particle filtering, and finally the extended Kalman filter and simultaneous localization and mapping.

Instead of focusing on state-of-the-art solutions to a particular sub-problem, the book's emphasis is on a progressive step-by-step development concepts through recurrent examples that capture the essence of a problem. The described solutions might not necessarily be the best, but they are easy to comprehend and widely used in the community. For example, odometry and line fitting are used to explain forward kinematics and least-squares solutions, respectively, and later serve as motivating examples for error propagation and the Kalman filter in a localization context.

Notably, the book is explicitly robot-agnostic, reflecting the timeliness of fundamental concepts. Rather, a series of possible project-based curricula are described in an appendix and available online, ranging from a maze-solving competition that can be realized with most camera-equipped differential-wheel robots to manipulation experiments with a robotic arm, all of which can be entirely conducted in simulation to teach most of the core concepts.

After multiple years of development and distribution mainly via Github, this new edition of the book has been coauthored with my colleagues in the Computer Science department—Bradley Hayes, Christoffer Heckman, and Alessandro Roncone, each having taught multiple iterations of our "Introduction to Robotics" and "Advanced Robotics" courses as well as special topics courses that pertain to their subfields of robotics research. They are adding not only tremendous technical depth but also years of experience on how certain subjects should be taught to remain engaging and exciting.

This book is released under a Creative Commons CC BY-NC-ND 4.0 International license, which allows anyone to copy and share its source code. However, neither the compiled version nor the code shall be used to create derivatives for commercial purposes. We have chosen this format as it seems to maintain the best trade-off between a freely available textbook resource that others may contribute to and a consistent curriculum that others can refer to. We are incredibly grateful to MIT Press and our editor Elizabeth Swayne for their support of this forward-looking model.

Writing this book would not have been possible without the excellent work of others before us, most notably *Introduction to Robotics: Mechanics and Control* by John Craig and *Introduction to Autonomous Mobile Robots* by Roland Siegwart, Illah Nourbakhsh, and Davide Scaramuzza, and innumerable other books and websites from which I learned and borrowed examples and notation. We are also grateful to Brian Amberg, Aaron Becker, Bachir El-Kadir, James Grime, Michael Sambol, Cyrill Stachniss, Subh83, and Ethan Tiran-Thompson, who made lecture video snippets and animations available online that are referenced throughout the book using QR codes.

I would like to acknowledge Mike Miles and Harel Biggie, graduate students in the authors' shared laboratory at the University of Colorado Boulder, for their careful reading and contributions. Finally, I would also like to acknowledge Github users AlWiVo, beardicus, mguida22, aokeson, as1ndu, apnorton, John Allen, jmodares, countsoduku, choffmann, and chrstphrdlz for their pull requests and comments, as well as Haluk Bayram. Your interest and motivation in this project has been one of our biggest rewards.

Nikolaus Correll
Boulder, Colorado, February 8, 2022

1 Introduction

Robotics celebrated its 60th birthday in 2021. The first commercial robot, the Unimate, made an appearance on a segment of *The Tonight Show* in 1966. At the time, this robot did amazing things: It opened a bottle of beer, poured a glass, put a golf ball into a hole, and even conducted an orchestra. It did all the things we expect a good robot to do. It was dexterous, accurate, and even creative. With all the technological advances that have occurred in the intervening decades, just how incredible must the capabilities of today's robots be and what must they be able to do?

Interestingly, we only recently developed the techniques to autonomously do all the things demonstrated by the Unimate. Unimate indeed did what was shown on TV, but all of its motions were preprogrammed and the environment was carefully staged. Only the advent of cheap and powerful sensors and the surge in computation capabilities have recently enabled robots to detect objects by themselves, plan motions to reach for them, and ultimately grasp and manipulate. Yet robotics is still far away from doing these tasks with humanlike performance.

This book introduces you to the computational fundamentals behind the design and control of autonomous robots. Robots are considered to be *autonomous* when they make decisions in response to their environment (rather than simply following a preprogrammed set of motions). They achieve this using a multitude of modern techniques ranging from signal processing, control theory, artificial intelligence, and more. These techniques are tightly intertwined with the mechanics, the sensors, and the actuators of the robot. Designing a robot therefore requires a deep understanding of both algorithms and its interfaces to the physical world.

The goals of this introductory chapter are to introduce the common problems that roboticists deal with and how they solve them.

1.1 Intelligence and Embodiment

Our notion of "intelligent behavior" is strongly biased by our understanding of the brain and how computers work: Intelligence is located in our heads. In fact, however, a lot of behavior that looks intelligent can be achieved by very simple mechanisms. For example,

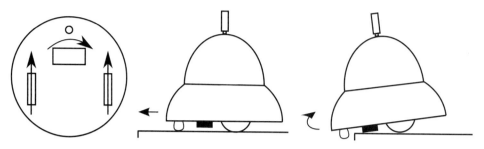

Figure 1.1
A windup toy that does not fall off the table using purely mechanical control. A flywheel that turns orthogonal to the robot's motion induces a right turn as soon as it hits the ground once the front caster wheel goes off the edge.

mechanical windup toys can avoid falling off an edge simply by using a flywheel that rotates at a right angle to their direction of motion and a caster wheel. Once the caster wheel loses contact with the ground—that is, when the robot has reached the edge—the flywheel kicks in and pulls the robot to the right (figure 1.1).

A robot vacuum cleaner might solve the same problem very differently: It employs infrared sensors that are pointed downward to detect edges such as those found on stairs and issues a command to make an avoiding turn in response. Given that onboard electronics are needed, this is a much more efficient, albeit more complex, approach.

Even though the above examples provide different approaches to implement intelligent behaviors, similar trade-offs exist for robotic planning. For example, ants can find the shortest path between their nest and a food source by simply choosing the trail that already has more pheromones (the chemicals ants communicate with) on it. Because ants not only moving faster toward the food but also return faster when taking the shorter paths, their pheromone trails build up quicker on the short paths (figure 1.2). But ants are not stuck to this solution. Every now and then, ants give the longer path another shot, eventually finding new food sources. What looks like intelligent behavior at the swarm level is essentially achieved by a pheromone sensor that occasionally fails. A modern industrial robot would solve the problem completely differently: it would first acquire some representation of the environment in the form of a map populated with obstacles and then plan a path using an algorithm.

Which solution to achieve a certain desired behavior is best depends on the resources that are available to the designer. We will now study a more elaborate problem for which many solutions, all more or less efficient, exist.

1.2 A Roboticists' Problem

Imagine the following scenario. You are a robot in a maze like environment such as a cluttered warehouse, hospital, or office building. There is a chest full of gold coins hidden somewhere inside. Unfortunately, you don't have a map of the maze. In the event you find

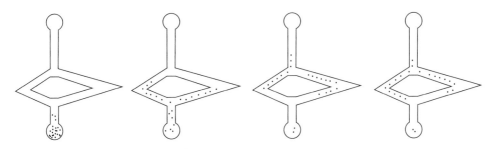

Figure 1.2
Ants finding the shortest path from their nest (bottom) to a food source (top). From left to right: The ants initially have equal preference for the left and the right branch, both going back and forth. As ants return faster on the shorter branch, there will be more pheromones present on the short branch once a new ant arrives from the nest.

the chest, you may only take a couple of coins at a time and bring them to the exit door where your car is parked.

Think about a strategy that will allow you to harvest as many coins in the shortest time as possible. Think about the cognitive and perception capabilities you would use. Now discuss alternative strategies: If you could not use these capabilities, what would you do? For example, what if you were blind or had no memory of the past?

These are exactly the same problems a robot has. A robot is a mobile machine that may reason about its environment with sensors and computation. Current robots are far from possessing the capabilities humans have, therefore it is worth considering what strategies *you* would employ to solve a problem if *you* were to lack some important perception or computational capabilities.

Before we move forward to discuss potential strategies for robots with impeded sensory systems, let's rely on a little bit on what we know from studying algorithms and briefly consider a particular strategy. You will need to explore the maze without entering any branch twice. You can use a technique known as *depth-first search* to do this, but you will need to be able to not only map the environment but also localize in the environment by recognizing places and by estimating locations by dead-reckoning on the map. Once you find the gold, you would need to plan the shortest path back to the exit, which you can then use to go back and forth until all the gold is harvested.

1.3 Ratslife: An Example of Autonomous Mobile Robotics

Ratslife is a miniature robot maze competition, developed by Olivier Michel from Cyberbotics S.A., which exemplifies a broad range of topics covered in this book. The Ratslife environment can easily be created from LEGO bricks, cardboard, or wood, and the game

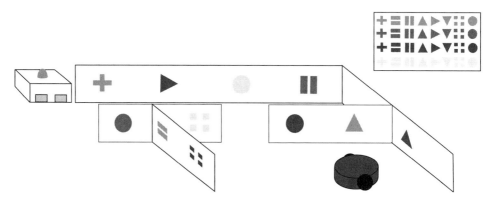

Figure 1.3
A simple maze made from cardboard, wood, or LEGO bricks with one or more charging stations. Locations in the maze are marked with unique markers that can be recognized by a simple robot.

can be played with any two mobile robots, preferably ones with the ability to identify markers in the environment. Examples are simple differential-wheel educational platforms with onboard cameras or even a smartphone-driven robot. Figure 1.3 shows an example environment that can be constructed from craft materials and illustrates some practical aspects of mobile robots for competitions.

In Ratslife, two miniature robots compete while searching for four "feeders" that are hidden in a maze. Once a robot reaches a feeder, it receives "energy" to go on for another 60 seconds, and the feeder becomes temporarily unavailable. After a short while, the feeder becomes available again. The feeders can be either controlled by a referee (who also takes care of time-keeping) or constructed as part of a simple curriculum on electronics or mechatronics.

It should be clear by now how *you* might solve these tasks using your abilities, and you may have also thought about some fallback strategies in case a sensor or two of yours were unavailable. Here are some possible algorithms for a robot, ordered by increasingly sophisticated capabilities the robot might wield:

• Imagine you have a robot that can only drive (actuation) and bounce off a wall. The resulting random walk will eventually let the robot reach a feeder. Because the allowed time to do so is limited, it is likely that the robot's energy will soon deplete.

• Now imagine a robot that has a sensor that gives it the ability to estimate its distance from a wall. This could be a whisker, an infrared distance sensor, an ultrasound distance sensor, or a laser range finder. The robot could now use this sensor to keep following a wall to its right. Using this strategy for solving the maze, it will eventually explore the entire maze except for islands inside of it.

• Finally, think about a robot that could identify simple patterns using vision, has distance sensors to avoid walls, and an "odometer" to keep track of its wheel rotations. Using these capabilities, the robot's potential winning strategy would be to explore the environment, identify markers in the environment using vision and then use them to create a map of all feeder locations, calculate the shortest path from feeder to feeder, and keep going back and forth between them. Strategy-wise, it might make sense to wait just in front of the feeder and approach it only shortly before the robot runs out of power.

1.4   Autonomous Mobile Robots: Some Core Challenges

Being able to stitch sensor information together to map the environment just by counting your own steps and orienting yourself by using distinct features of the environment is known as simultaneous localization and mapping (SLAM). The key challenge here is that the length of the steps you take are uncertain (a wheeled robot might slip or have wheels of slightly different size), and it is not possible to recognize places with 100 percent accuracy (even humans can have trouble with this). In order to be able to implement something like the last algorithm on a real robot, we will therefore need to understand the answers to the following questions:

- How does a robot move? How does the rotation of its wheels affect its position and speed in the world?
- How might we to control the wheel speed in order to reach a desired position?
- What sensors exist for a robot to perceive its own status and its environment?
- How can we extract structured information (e.g., features of the world) from this vast amount of sensor data?
- How can we localize in the world?
- How can error be represented, and how can we reason in the face of uncertainty?

To answer these questions, we will rely on trigonometry, calculus, linear algebra, probability, and algorithms. Specific concepts that will be used throughout this book are basic trigonometry, derivatives and integrals, matrix notation, Bayes' formula, and the concept of probability distributions. Robotics is a great vehicle to add meaning to these concepts!

1.5   Autonomous Manipulation: Some Core Challenges

Think about the last time you worked with your hands: typing on your keyboard, writing on a piece of paper, sewing a button onto a shirt, and using a hammer or a screwdriver. You will notice that these activities require a wide range of dexterity (i.e., the ability to manipulate objects with precision), a wide range of forces, and a wide range of sensorial capabilities. You will also notice that some tasks go beyond your natural capabilities, such as putting yarn through a hole in fabric, turning a screw, or driving a nail into a piece of wood, but they can be easily solved with the right tool.

Present-day robotic hands are far from reaching the dexterity of a human hand. Yet, with the right tool (called the "end-effector" in robotics speech), some tasks can be solved even faster and more precisely by a robot equipped with the right tool than by a capable human. Just as with solving a mobile robotics problem, manipulation problems require you to think about the right mix of reasoning and mechanism design. For example, grasping tiny parts might be impossible with tweezers but quite easy when employing a sucking mechanism. Or picking up a test tube that is nearly invisible can be picked up almost blindly when using a funnel-like mechanism at your end-effector. Unfortunately, these tricks will most likely limit the versatility of your robot, requiring you to think about the problem and the users' needs as a whole.

Take-Home Lessons

• The best solution to a problem is a function of the available sensing, actuation, computation, and communication abilities of the available platform. Usually, there are trade-offs that allow you to solve a problem using a minimal set of resources but that compromise performance characteristics such as speed, accuracy or reliability.

• Robotics problems are different from many problems in pure artificial intelligence, particularly those that do not deal with unreliable sensing or actuation.

• The unreliability of sensors, actuators, and communication links requires a probabilistic notion of the system and the ability to reason with uncertainty.

Exercises

1. What kind of sensors do you need to solve the Ratslife game? Think both about trivial and close-to-optimal approaches.

2. What devices in your home could be considered robots? Why or why not?

3. Is a mechanical clock a robot? Why or why not?

4. Which industries have been recently revolutionized by robotics? Which industries were the first to introduce robots? Which industries are currently being transformed?

5. What sensors are you using when you grasp an object? Enumerate them all. Which ones are absolutely necessary and which could you live without?

6. Think about robots vacuuming your floor or mowing your lawn. Do they use any planning? Is planning necessary? Why or why not?

7. What kind of sensors would you need in a car that drives completely autonomously? Think first about the kind of information that the car needs to be aware of and then discuss possible sensors that could capture this information.

8. Implement a simple line-following exercise using a robot of your choice. How does the thickness of the line affect the sensor placement on the robot? How does its curvature affect the robot's maximum speed?

9. Implement a maze-solving algorithm that uses simple wall-following instructions and using a robot of your choice. How does the sensor geometry affect the robot's performance? What are the parameters that you find yourself tuning?

I MECHANISMS

2 Locomotion, Manipulation, and Their Representations

Autonomous robots are systems that sense, compute, communicate, and actuate. Actuation, the focus of this chapter, is the ability of the robot to move and to manipulate the world. Specifically, we differentiate between locomotion as the robot's ability to move itself and manipulation as the robot's ability to move objects in the environment. Both activities are closely related: During locomotion the robot uses its motors to exert forces on its environment (ground, water, or air) to move itself; during manipulation it uses motors to exert forces on objects to move them relative to the rest of the environment. This might not even require different motors. Insects are good examples because they can use their six legs not only for locomotion but also for picking up and manipulating objects. In short, the goals of this chapter are to:

- Introduce the concepts of locomotion, manipulation, and their duality.
- Explain static versus dynamic stability.
- Introduce the concept of "degree of freedom" (DoF).
- Introduce coordinate systems and their transformations.

2.1 Locomotion and Manipulation Examples

Locomotion includes very different concepts of motion, including rolling, walking, running, jumping, sliding (undulatory locomotion), crawling, climbing, swimming, and flying. The mechanisms that might achieve these feats could be drastically different in terms of energy consumption, kinematics, stability, and other capabilities required by the robot that implements them. Furthermore, the definitions are loose and ambiguous. For example, "swimming" can be performed using many different forms of propulsion. Similarly, a sliding motion on the ground might work well for swimming too, with only a few modifications.

The way in which the individual parts of a robot can move with respect to each other and the environment is called the *kinematics* of the robot. Kinematics (discussed in detail

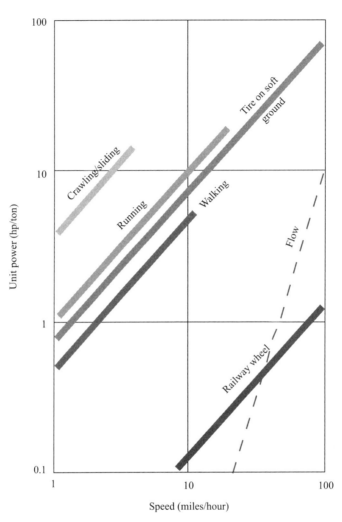

Figure 2.1
Power consumption versus speed for various means of locomotion. *Source*: Siegwart (2011) and Todd (1985).

in chapter 3) are only concerned with the position and speed (first derivative of position) of those parts; depending on the application, one may want to use a deeper level of abstraction called *dynamics*, which is concerned with quantities such as acceleration (second derivative of position) and jerk (third derivative of position).

Commercially, the most widespread form of locomotion is rolling. This is partially because rolling provides by far the most efficient energy-to-speed ratio (see figure 2.1), making the invention of the wheel one of the greatest technological breakthroughs in history. It also is a widely implemented form of locomotion because of cars and bicycles.

Consequently, humans have modified their environment to have as many smooth surfaces as possible—for example, roads and floors in buildings. In contrast, evolution has not equipped any animal with wheel-like actuators because of their poor performance in natural environments—consider, for example, an unmown meadow, a forest floor, a mountain or a cave. Consequently, wheeled robots perform poorly in such environments whereas legged robots can shine.

Can you find examples of robots from the above categories (legged versus wheeled robots)? Identify the different types of actuators that are used in them.

Most mechanisms capable of locomotion can also be used for manipulation with only minor modifications. Most industrial manipulators consist of a chain of rotary (or revolute) actuators that are connected by rigid links. In general, they are equipped with six or more independently rotating axes—we will see why later. In addition, modern industrial manipulators have the ability to not only control the position of each joint but to also control the torque at each individual joint; this capability allows control over the *compliance* of a robot, which in a mechanical sense is the inverse of stiffness. Finally, for dexterous manipulation, a robot does not only need an arm but also a gripper or hand. Grasping is a hard problem on its own and is therefore treated in its own chapter (chapter 5).

Regardless of whether the robot is rolling or walking, the dominant actuator type is rotational. Another type of mechanism is the *prismatic* or *linear* joint (see figure 2.5) that allows the robot to extend and contract a link. Joints of this type are usually combined with rotating joints and allow, for example, a robot arm to move up and down or a robotic walker to extend or retract its leg.

2.2   Static and Dynamic Stability

A fundamental difference between locomotion mechanisms is whether they are statically or dynamically stable. A statically stable mechanism will not fall even when not actuated (figure 2.2, left). A dynamically stable robot instead requires constant actuation to prevent it from falling. Technically, stability requires the robot to keep its center of mass to fall within the polygon spanned by its ground-contact points. For example, a quadrupedal robot's feet span a rectangle. Once such a robot lifts one of its feet, this rectangle becomes a triangle. If the projection of the center of mass of the robot along the direction of gravity is outside this triangle, the robot will fall. A dynamically stable robot can overcome this problem by changing its configuration so rapidly that a fall is prevented. An example of a purely dynamically stable robot is an inverted pendulum on a cart (figure 2.2, middle). Such a robot has no statically stable configurations and needs to keep moving all the time to keep

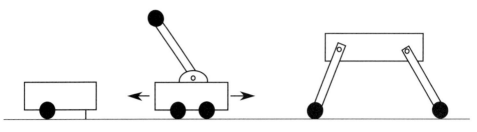

Figure 2.2
From left to right: Statically stable robot; dynamically stable inverted pendulum robot; static and dynamically stable robot (depending on configuration).

the pendulum upright. While dynamic stability is desirable for high-speed, agile motions, robots should be designed so that they can easily switch to a statically stable configuration (figure 2.2, right).

An example of a robot that has both statically and dynamically stable configurations is a quadruped running. Unlike walking, running requires the quadruped robot to always have two legs in the air and alternate between them fast enough before the robot has a chance to fall. Although statically stable walking is possible with only four legs, most animals (and robots) require six legs for statically stable walking, and they use dynamically stable gaits (such as galloping) when they have four legs. Six legs allow the animal to move three legs at a time while the three other legs maintain a stable pose.

2.3   Degrees of Freedom

The concept of *degree of freedom*, often abbreviated as DoF, is important for defining the possible positions and orientations a robot can reach. An object in the physical world can have up to six *Cartesian* degrees of freedom—namely, forward/backward, sideways, and up/down as well as rotations around those axes. These rotations are known as pitch, yaw, and roll and are illustrated in figure 2.3. These Cartesian degrees of freedom are distinct from the robot's *mechanical* degrees of freedom, which correspond to the number of points of actuation for a robot (i.e., a robotic arm with five joint motors is referred to as having five mechanical degrees of freedom *in joint space*, which is discussed in chapter 3). As a rule of thumb, the number of mechanical DoFs available to the user depends on the robot platform and cannot easily be changed by the user unless mechanical modifications to the robot are made; conversely, the number of Cartesian DoFs depends on the task, can be modified by the user, and varies according to what the robot needs to do.

After specifying the mechanical and Cartesian DoFs for your kinematic problem, the number of Cartesian DoFs (i.e., directions) a robot can actually move in depends on the configuration of its actuators and the constraints the robot has with the environment. These relationships are not always intuitive and require more rigorous mathematical treatment

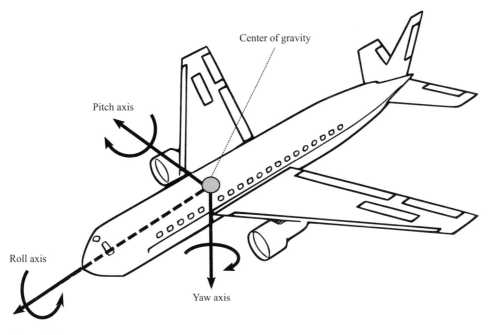

Center of gravity

Pitch axis

Roll axis

Yaw axis

Figure 2.3
Pitch, yaw, and roll around the principal axis of an airplane.

(see chapter 3). The goal of this section is to introduce the degrees of freedom of standard mechanisms that are recurrent in robot design, such as wheels or simple arms. For wheeled platforms, the degrees of freedom are defined by the types of wheels used and their orientation. Common wheel types are listed in table 2.1.

Only robots that exclusively use wheels with three degrees of freedom (3-DoF wheels) will be able to freely move on a plane. This is because the pose of a robot on a plane is fully determined by its position (two values, which are vertical and horizontal position) and its orientation (one value such as an angle). Robots that don't have wheels with three degrees of freedom will have *kinematic constraints* that prevent them from reaching every possible point at every possible orientation. For example, a bicycle wheel can only roll along one direction and turn on the spot. Moving the bicycle wheel orthogonally to its direction of motion is not possible unless it is forcefully dragged ("skidding"). Importantly, not having three degrees of freedom does not imply that some poses in the plane are unreachable—it may just require additional movements to achieve them!

A good analogue are figures on a chessboard. For example, a knight can reach every cell on a chessboard but might require multiple moves to do so. This is similar to a car, which can parallel park using back-and-forth motions. Instead, a bishop can only reach either black or white fields on the board, based on its starting position.

Table 2.1
Different types of wheels and their degrees of freedom.

Wheel type	Example	Degrees of freedom
Standard	Front wheel of a wheelbarrow	Two: • Rotation around the wheel axle • Rotation around its contact point with the ground
Caster wheel	Office chair	Three: • Rotation around the wheel axle • Rotation around its contact point with the ground • Rotation around the caster axis
Swedish wheel	Standard wheel with non-actuated rollers around its circumference	Three: • Rotation around the wheel axle • Rotation around its contact point with the ground • Rotation around the roller axles
Spherical wheel	Ball bearing	Three: • Rotation in any direction • Rotation around its contact point

Source: Adapted from Siegwart, Nourbakhsh, and Scaramuzza (2011, p. 36).

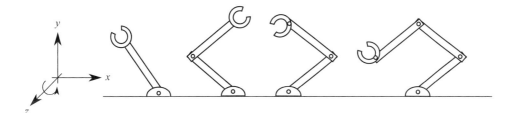

Figure 2.4
From left to right: Manipulators with one, two, three, and four mechanical DoFs. The Cartesian DoFs needed for the end-effector to move in a plane are the vertical displacement of the end-effector with respect to the base, its horizontal displacement, and its orientation.

Similar reasoning applies to aerial and underwater robots. Here, the position of the robot is affected by the position and orientation of its thrusters, either in the form of jets or propellers. Things become complicated quickly, however, as the dynamics of the system are subject to fluid-dynamic and aerodynamic effects, which also change as a function of the size of the robot. This book does not go into the details of flying and swimming robots, but the general principles of localization and planning are applicable to them as well.

> Think about possible wheel, propeller, and thruster configurations. Don't limit yourself to robots. Consider also street and aerial vehicles and be creative—if you can think about a setup that makes sense and that allows for reasonable mobility, then somebody has already built it and analyzed it. What are the advantages and disadvantages of each?

For manipulating arms, Cartesian DoFs refer to the positions and orientations—rotations around the primary axes (x, y, and z)—that the end-effector can reach. Each actuated joint will typically add a degree of freedom unless it is redundant (moving in the same direction, with the same physical effect, as a different joint). Figures 2.4 and 2.5 show a series of manipulators operating on a planar surface. In such a scenario, the degrees of freedom of the end-effector are limited to moving up and down, sideways, and rotating around their pivot point. As a plane only has those three degrees of freedom, adding additional joints will not increase the number of Cartesian DoFs unless they allow the robot to also move in and out of the plane ("vertical" axis). An exact definition of the number of degrees of freedom is tricky and requires deriving analytical expressions for the end-effector position and orientation, which is the subject of chapter 3.

Choosing the "right" kinematics involves a very complex trade-off between mechanical complexity, maneuverability, achievable precision, cost, and ease of control. The very popular differential-wheel drive—consisting of two independently controlled wheels that

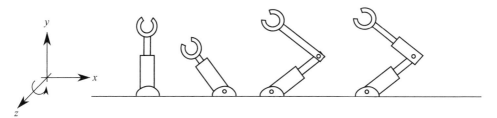

Figure 2.5
From left to right: Manipulators with one, two, three, and four DoFs using a combination of rotational and prismatic joints.

share a common axis, such as those mounted on a robotic vacuum cleaner—is cheap,
highly maneuverable, and easy to control; however, it is hard to drive the robot in a perfectly straight line. This motion requires both motors to turn at the exact same speed and
both wheels to have the exact same diameter, which is hard to achieve in practice. This
problem is solved well by carlike steering mechanisms—which in turn have poor maneuverability and are difficult to control (as a reference, think about the complexity of parallel
parking).

2.4   Coordinate Systems and Frames of Reference

Every robot assumes a *pose* in the real world that can be described by its position (x, y,
and z) and orientation (pitch, yaw, and roll) along the three major axes of a Cartesian coordinate system. Such a coordinate system is shown in figure 2.6. Note that the directions
and orientations of the coordinate axes are arbitrary. This book uses the "right-hand rule,"
which is illustrated in figure 2.6 to determine axes labels and directions throughout. Pitch,
yaw, and roll are also known as bank, attitude, and heading in other communities. This
makes sense, considering the colloquial use of the word "heading," which corresponds to a
rotation around the z-axis of a vehicle driving on the x-y-plane.

   Defining all three position axes and orientations might be cumbersome. What level of
detail we care about, where the origin of this coordinate system is, and even what kind of
coordinate system we choose will depend on the specific application. For example, a simple
mobile robot would typically require a representation with respect to a room, a building, or
the earth's coordinate system (given by the longitude and latitude of each point on earth),
whereas a static manipulator usually has the origin of its coordinate system at its base. More
complicated systems, such as mobile manipulators or multilegged robots, make life much
easier by defining multiple coordinate systems (e.g., one for each leg and one that describes
the position of the robot in the world frame). These local coordinate systems are known as
frames of reference. An example of two nested coordinate systems is shown in figure 2.7.

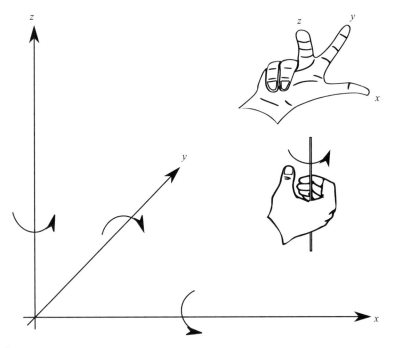

Figure 2.6
A coordinate system indicating the direction of the coordinate axes and rotation around them. These directions have been derived using the right-hand rules.

In this example, a robot located at the origin of x', y', and z' might plan its motions in its own reference frame, which can then be expressed in the coordinate system x, y, and z by performing a translation and a rotation—as we will later see.

Depending on its degrees of freedom in Cartesian space—that is, the number of independent translations and rotations a robot can achieve in such a space—it is also customary to ignore components of position and orientation that remain constant. For example, a simple floor-cleaning robot's pose might be completely defined by its x and y coordinates in a room as well as its orientation (i.e., its rotation around the z-axis). In this case, z position and rotation around x and y axes would be ignored.

2.4.1 Matrix Notation

Given some kind of fixed coordinate system, we can describe the *position* of a robot's end-effector by a 3×1 position vector. As there can be many coordinate systems that defined on a robot and the environment, we identify the coordinate system that a point relates to by a preceeding superscript—for example, ^{A}P to indicate that point P is in coordinate system $\{A\}$. Each point consists of three elements: $^{A}P = [p_x, p_y, p_z]^T$.

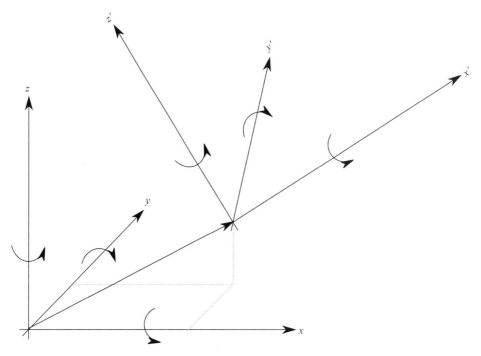

Figure 2.7
Two nested coordinate systems (also referred to as frames of reference).

More formally, ^AP is a linear combination of the three basis vectors that span A:

$$^AP = p_x \begin{bmatrix} 1 \\ 0 \\ 0 \end{bmatrix} + p_y \begin{bmatrix} 0 \\ 1 \\ 0 \end{bmatrix} + p_z \begin{bmatrix} 0 \\ 0 \\ 1 \end{bmatrix}. \tag{2.1}$$

As we know, not only the position of the robot is important but also its orientation. In order to describe the orientation of a point, we will attach a coordinate system to it. Let $\hat{X}_B$, $\hat{Y}_B$ and $\hat{Z}_B$ be unit vectors that correspond to the principal axes of a coordinate system $\{B\}$. When expressed in coordinate system $\{A\}$, they are denoted $^A\hat{X}_B$, $^A\hat{Y}_B$ and $^A\hat{Z}_B$. In order to express a vector that is given in one coordinate system in another, we need to *project* each of its components to the unit vectors that span the target coordinate system. For example, let's say we consider only the axis $^A\hat{X}_B$, which is given by

$$^A\hat{X}_B = (\hat{X}_B \cdot \hat{X}_A, \hat{X}_B \cdot \hat{Y}_A, \hat{X}_B \cdot \hat{Z}_A)^T, \tag{2.2}$$

that is, the projections of $\hat{X}_B$ onto $\hat{X}_A$, $\hat{Y}_A$, and $\hat{Z}_A$. Here, the '·' denotes the scalar product, also known as dot or inner product (see appendix B.1). Note that all vectors in (2.2) are unit vectors—that is, their length is one. By following the definition of the scalar product, we have $A \cdot B = \|A\| \|B\| \cos \alpha = \cos \alpha$, which indeed reduces the projection of $\hat{X}_B$ onto the unit vectors of $\{A\}$. This projection is illustrated in figure 2.8.

We can now apply the same procedure to all three vectors that span coordinate system $\{B\}$ and stack these three vectors together into a 3×3 matrix to obtain the rotation matrix

$$_B^A R = [^A\hat{X}_B \quad ^A\hat{Y}_B \quad ^A\hat{Z}_B], \tag{2.3}$$

which describes $\{B\}$ relative to $\{A\}$. It is important to note that all columns in $_B^A R$ are unit vectors, so the rotation matrix is orthonormal. This is important because it allows us to easily obtain the inverse of $_B^A R$ as $_B^A R^T$ or $_A^B R = _B^A R^T$.

The reason why the unit vectors of a coordinate system $\{B\}$ expressed in coordinate system $\{A\}$ actually make up a rotation matrix can be easily seen when rearranging equation (2.1) in matrix form:

$$^A P = \begin{bmatrix} 1 & 0 & 0 \\ 0 & 1 & 0 \\ 0 & 0 & 1 \end{bmatrix} \begin{bmatrix} p_x \\ p_y \\ p_z \end{bmatrix}, \tag{2.4}$$

where the rotation matrix is the identity as both points already are in the same coordinate system—that is, no rotation is needed.

We have now established how to express the orientation of a coordinate system using a rotation matrix. Usually, coordinate systems do not lie on top of each other but are also displaced from each other. Together, position and orientation are known as a *frame*, which is a set of four vectors, one for the position and three for the orientation, and we can write

$$\{B\} = \{_B^A R, ^A P\} \tag{2.5}$$

to describe the coordinate frame $\{B\}$ with respect to $\{A\}$ using a vector $^A P$ and a rotation matrix $_B^A R$. Robots usually have many such frames defined along their bodies.

2.4.2 Mapping from One Frame to Another

Having introduced the concept of frames, we need the ability to map coordinates in one frame to coordinates in another frame. For example, let's consider frame $\{B\}$ having the same orientation as frame $\{A\}$ and sitting at location $^A P$ in space. As the orientation of both frames is the same, we can express a point $^B Q$ in frame $\{A\}$ as

$$^A Q = ^B Q + ^A P. \tag{2.6}$$

In reality, adding two vectors that are in different reference frames (i.e., $^B Q + ^A P$) is only possible if both of them have the same orientation. We can, however, convert from one reference frame to the other using the rotation matrix

$$^A P = _B^A R {}^B P \tag{2.7}$$

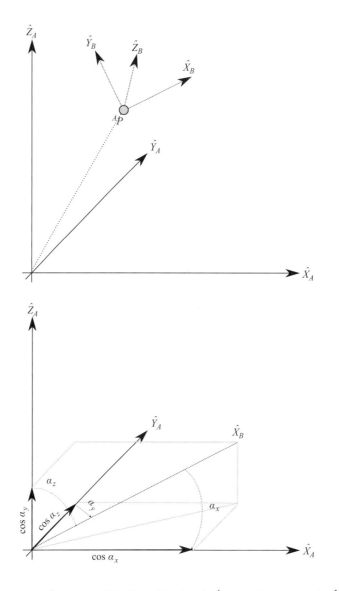

Figure 2.8
The top graph shows a coordinate system {B} with position given by ^AP and orientation given by $\hat{X}_B$, $\hat{Y}_B$, and $\hat{Z}_B$. The bottom graph shows the projection of the unit vector $\hat{X}_B$ onto the unit vectors that span coordinate system {A} after moving {B} into the origin of {A}. As all vectors are unit vectors, $A \cdot B = \|A\|\|B\| \cos\alpha = \cos\alpha$.

and therefore solve the mapping problem regardless of the orientation of $\{A\}$ to $\{B\}$:

$$^AQ = {}^A_B R\, {}^B Q + {}^A P. \tag{2.8}$$

Using this notation, we can see that leading subscripts cancel the leading superscripts of the following vector/rotation matrix. Even though we have now a solution to transform a point from one frame of reference to another by combining a rotation and a translation, it would be more appealing to write it in a more compact form, namely:

$$^AQ = {}^A_B T\, {}^B Q. \tag{2.9}$$

In order to do this, we need to introduce a 4×1 position vector such that

$$\begin{bmatrix} {}^A Q \\ 1 \end{bmatrix} = \begin{bmatrix} {}^A_B R & {}^A P \\ 0 \ 0 \ 0 & 1 \end{bmatrix} \begin{bmatrix} {}^B Q \\ 1 \end{bmatrix} \tag{2.10}$$

and $^A_B T$ is a 4×4 matrix. Note that the added 1s and [0 0 0] do not affect the other entries in the matrix during matrix multiplication. A 4×4 matrix of this form is called a *homogeneous transform*.

The inverse of a homogeneous transform can be constructed by inverting the rotation and translation part independently, leading to

$$\begin{bmatrix} {}^A_B R & {}^A P \\ 0 \ 0 \ 0 & 1 \end{bmatrix}^{-1} = \begin{bmatrix} {}^A_B R^T & -{}^A_B R^T\, {}^A P \\ 0 \ 0 \ 0 & 1 \end{bmatrix}. \tag{2.11}$$

We have now established a convenient notation to convert points from one coordinate system to another. There are many possible ways this can be done, in particular how rotation can be represented (see below), but all can be converted from one form into the other.

2.4.3 Concatenation of Transformations

Transformations can be combined. Consider, for example, an arm with two links, reference frame $\{A\}$ at the base, $\{B\}$ at its first joint, and $\{C\}$ at its end-effector. Given the transforms $^B_C T$ and $^A_B T$, we can write

$$^A P = {}^A_B T\, {}^B_C T\, {}^C P = {}^A_C T\, {}^C P \tag{2.12}$$

to convert a point in the reference frame of the end-effector to that of its base. As this works for rotation and translation operators independently, we can construct $^A_C T$ as

$$^A_C T = \begin{bmatrix} {}^A_B R\, {}^B_C R & {}^A_B R\, {}^B P_C + {}^A P_B \\ 0 \quad 0 \quad 0 & 1 \end{bmatrix}, \tag{2.13}$$

where $^A P_B$ and $^B P_C$ are the translations from $\{A\}$ to $\{B\}$ and from $\{B\}$ to $\{C\}$, respectively.

2.4.4 Other Representations for Orientation

So far, we have represented orientation by a 3×3 matrix whose column vectors are orthogononal unit vectors describing the orientation of a coordinate system. Orientation is therefore represented with nine different values. We chose this representation mainly because it is the most intuitive to explain and is derived from simple geometry.

Euler angles

In fact, three values are sufficient to describe orientation. This becomes clear when considering that orthogonality (dot product of all columns is zero) and vector length (each vector must have length 1) impose six constraints on the nine values in the rotation matrix. Indeed, an orientation can be represented as a rotation by certain angles around the x, the y, and the z-axis of the reference coordinate system. This is known as the X-Y-Z fixed-angle notation. Mathematically, this can be represented by a rotation matrix of the form:

$$
{}_{B}^{A}R_{XYZ}(\gamma, \beta, \alpha) = \begin{bmatrix} \cos\alpha & -\sin\alpha & 0 \\ \sin\alpha & \cos\alpha & 0 \\ 0 & 0 & 1 \end{bmatrix} \begin{bmatrix} \cos\beta & 0 & \sin\beta \\ 0 & 1 & 0 \\ -\sin\beta & 0 & \cos\beta \end{bmatrix} \begin{bmatrix} 1 & 0 & 0 \\ 0 & \cos\gamma & -\sin\gamma \\ 0 & \sin\gamma & \cos\gamma \end{bmatrix}.
$$

$$(2.14)$$

While the X-Y-Z fixed-angle approach expresses a coordinate frame using rotations with respect to the original coordinate frame (say, $\{A\}$), another possible description is to start with a coordinate frame $\{B\}$ that is coincident with frame $\{A\}$, then rotate around the Z-axis with angle α, then the Y-axis with angle β, and finally around the X-axis with angle γ. This representation is called Z-Y-X Euler angles. As the coordinate axis do not necessarily need to be different, there are 12 possible valid combinations of subsequent rotations: XYX, XZX, YXY, YZY, ZXZ, ZYZ, XYZ, XZY, YZX, YXZ, ZXY, and ZYX. The reason there are only 12 combinations is that subsequent rotations around the same axis are not valid. Such rotations would not add any information; they are equivalent to a rotation by the sum of both angles.

It is important to understand the subtle differences between the available transformations because there is neither a "right" nor a "wrong" convention. However, different manufacturers of hardware and software products might use different ones, often based on preferences in the different fields, such as aviation or geology, that these algorithms and products originally catered to. There is a fundamental caveat with all of the aforementioned approaches: Each of the rotation matrices can look like subsequent rotations around the same axis for certain values of angles. For example, this happens for the XYZ rotation matrix if the angle of rotation around the Y-axis is 90 degrees. These cases are known as a *singularities* of the specific notation. We therefore need additional representations that work across the entire range of possible motions.

Quaternions

Among the many possible conventions, the preferred representation for computational and stability reasons are *quaternions*. A quaternion is a four tuple (4-tuple) that extends the complex numbers with very general applications in mathematics and represents orientation and rotation in particular. Quaternions are generally represented in the form

$$q = a + b\mathbf{i} + c\mathbf{j} + d\mathbf{k}. \tag{2.15}$$

Here, a is referred to the *scalar* part of the quaternion and the elements b, c, and d as the *vector* part.

A quaternion's *conjugate* is given by

$$q^* = a - b\mathbf{i} - c\mathbf{j} - d\mathbf{k}. \tag{2.16}$$

It can be shown that each rotation can be represented as a rotation around a single axis (a vector in space) by a specific angle, also known as the Euler axis. Given such an axis $\hat{K} = [k_x k_y k_z]^T$ and an angle θ, one can calculate the so-called Euler parameters or unit quaternion $\mathbf{q} = (\epsilon_1, \epsilon_2, \epsilon_3, \epsilon_4)$ with

$$\epsilon_1 = \cos\frac{\theta}{2} \tag{2.17}$$

$$\epsilon_2 = k_x sin\frac{\theta}{2} \tag{2.18}$$

$$\epsilon_3 = k_y sin\frac{\theta}{2} \tag{2.19}$$

$$\epsilon_4 = k_z sin\frac{\theta}{2}. \tag{2.20}$$

These four quantities are constrained by the relationship

$$\epsilon_1^2 + \epsilon_2^2 + \epsilon_3^2 + \epsilon_4^2 = 1, \tag{2.21}$$

which might be visualized by a point on a unit hypersphere.

Given a vector $\mathbf{p} \in \mathbb{R}^3$ that should be rotated by a unit quaternion $\mathbf{q}$, we can compute the new vector $\mathbf{p}'$ as

$$\mathbf{p}' = \mathbf{q}\mathbf{p}\mathbf{q}^*, \tag{2.22}$$

with $\mathbf{q}^*$ the conjugate of $\mathbf{q}$ as defined above.

Computing the equivalent rotation to two subsequent rotations requires multiplying the quaternions. Given two quaternions ϵ and ϵ', multiplication is defined by the following matrix multiplication:

$$\begin{bmatrix} \epsilon_4 & \epsilon_1 & \epsilon_2 & \epsilon_3 \\ -\epsilon_1 & \epsilon_4 & -\epsilon_3 & \epsilon_2 \\ -\epsilon_2 & \epsilon_3 & \epsilon_4 & -\epsilon_1 \\ -\epsilon_3 & -\epsilon_2 & \epsilon_1 & \epsilon_4 \end{bmatrix} \begin{bmatrix} \epsilon_4' \\ \epsilon_1' \\ \epsilon_2' \\ \epsilon_3' \end{bmatrix}. \tag{2.23}$$

Unlike multiplying two rotation matrices, which requires 27 multiplications and 18 additions, multiplying two quaternions only requires 16 multiplications and 12 additions, making the operation computationally more efficient. Importantly, this representation does not suffer from singularities for specific joint angles, making the approach computationally more robust. This is particularly relevant for robotics, as mathematical singularities have a pretty significant real-world impact on physical robots.

Take-Home Lessons

• To perform planning for a robot, it is necessary to understand how its control parameters map to actions in the physical world.

• The kinematics of a robot are fully defined by the position and orientation of its wheels, joints, and links, no matter whether it swims, flies, crawls, or drives.

• Many robotic systems cannot be fully understood by considering kinematics alone; rather, they require you to model their dynamics as well. This book will be limited to modeling kinematics, which is sufficient for low-speed mobile robots and arms.

Exercises

1. What are the Cartesian DoFs of a push lawn mower with four wheels? How is it still possible to mow an entire lawn with one, even though the wheels don't yaw?

2. Is a car statically or dynamically stable? What about a motorcycle?

3. What are the Cartesian DoFs of an office chair with all caster wheels?

4. What are the maximum Cartesian DoFs for orientable objects driving on the two-dimensional plane?

5. What are the maximum Cartesian DoFs for objects that can freely translate and rotate in the world?

6. Calculate the Cartesian DoFs of a differential-drive robot with two powered rear wheels and a central, front-mounted caster wheel. What happens when you add a second caster wheel?

7. Calculate the Cartesian DoFs of a standard car. How is it possible to still reach every point on the plane?

8. A steering wheel allows you to change the yaw of your car. Can you also change its pitch and its roll? See figure 2.3 for reference.

3 Kinematics

In order to plan a robot's movements, we have to understand the relationship between our control variables (i.e., the input to the motors that we can control at any given time) and the effect of these control variables on the motion of the robot. The simplest models of such relationships can be built by looking at the geometry of our robot, known as the field of *kinematics*. For simple arms in static configurations, a kinematic model is rather straightforward: If we know the generalized position/configuration angle of each joint, we can calculate the generalized position of its end-effectors using trigonometry—a process known as *forward kinematics*. This process is usually more involved for mobile robots, as the speeds of the wheels need to be integrated to determine changes in robot pose, which we refer to as *odometry*. Roboticists are often concerned with trying to compute the inverse relationship: the position each joint must be at for the end-effector to be in a desired position or pose. This is generally a far more complex, underdetermined problem, known as *inverse kinematics*.

As we will see below, kinematics is the simplest and most fundamental level of abstraction that a roboticist can use to model the motion of a robot and its geometry. It deals exclusively with positional quantities and considers the robot as if it was frozen in time. Although this simplification is far from being realistic, we will see that a lot can be done through kinematics alone! However, a more expressive tool at our disposal is to do a similar modeling in a second level of abstraction that operates in velocity space; this domain, called *differential kinematics*, is introduced in section 3.3. In all, the goals of this chapter are to:

- Introduce the forward kinematics of simple arms and mobile robots, and understand the concept of holonomy.
- Provide an intuition on the relationship between inverse kinematics and path planning.
- Become familiar with differential kinematics and the Jacobian technique.

Within the scope of a kinematic analysis, the term *generalized position* or *generalized configuration* means "any position-equivalent quantity needed to describe the element." For what concerns joint space, it depends on the type of actuation: A revolute joint imparts a rotational motion around its axis and its configuration is fully described by an angle; a prismatic joint commands a translational motion along its axis and its configuration is represented by a distance. Conversely, generalized position in task space depends on the specific task; in its most general case, a generalized position equates to the end-effector pose, which consists of a three-dimensional (3D) position and a 3D orientation—as we will see below.

> Remember: Configuration space ≡ joint space; Cartesian space ≡ task space. Forward kinematics maps from joint space to task space, and inverse kinematics does the opposite. The number of mechanical degrees of freedom n (i.e., DoFs in task space) depends on the robot, while the number of Cartesian degrees of freedom m (i.e., DoFs in task space) depends on the task. In general, $n \neq m$!

3.1 Forward Kinematics

Now that we have introduced the notion of local coordinate frames, we are interested in calculating the pose and speed of these coordinate frames as a function of the robot's actuators and joint configuration. That is, we are interested in computing a function f that allows us to map a joint configuration to its corresponding end-effector pose:

$$r = f(q), \qquad f : \mathbb{R}^n \to \mathbb{R}^m, \tag{3.1}$$

where r is the task-space (end-effector) configuration and q is the joint-space configuration. It is important to remember that the choice of q and r (and, consequently, the complexity of f) depends on your specific robot platform and the specific task you are investigating. Generally, q refers to the actuators/joints that you can control on your robot; it is of size n, where n is the number of degrees of freedom (DoFs) in joint space (see also section 2.3). Conversely, r depends on the task and its dimensionality is m, where m is the number of DoFs in task space.

We will focus on the problem of computing the forward kinematics by mapping f for a variety of robot arms to build intuition. It is always possible to compute the forward kinematics *analytically* by inspecting the arm mechanism and the relationship between joint and task configuration (see section 3.1.1). However, in section 3.1.2 we will introduce a scalable, *geometric* technique to compute forward kinematics with more complex arms composed of many mechanical degrees of freedom.

3.1.1 Forward Kinematics of a Simple Robot Arm

Consider the robot arm in figure 3.1; it is mounted to a table, and is composed of two links and two joints. Let the length of the first link be l_1 and the length of the second link be l_2.

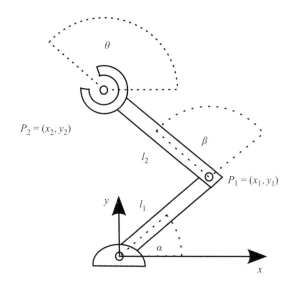

Figure 3.1
A simple 2-DoF arm.

You could specify the position of the link closer to the table by the angle α and the angle of the second link relative to the first link using the angle β. Therefore, $q = [\alpha, \beta]^T$ specifies the two degrees of freedom that we can control. Our goal is to calculate the position $[x, y]^T$ and orientation θ of the end-effector given the values of q; consequently, f will map to $r = [x, y, \theta]^T$.

Let's now calculate the position $P_1 = (x_1, y_1)$ of the joint between the first and the second link using simple trigonometry:

$$x_1 = l_1 \cos \alpha$$

$$y_1 = l_1 \sin \alpha. \tag{3.2}$$

Similarly, the position of the end-effector $P_2 = (x_2, y_2)$ is given by

$$x_2 = x_1 + l_2 \cos(\alpha + \beta)$$

$$y_2 = y_1 + l_2 \sin(\alpha + \beta). \tag{3.3}$$

For what concerns the orientation of the arm's end-effector θ, we know it is just the sum of $\alpha + \beta$. Altogether, the configuration r of the end-effector is given by

$$x = l_1 \cos \alpha + l_2 \cos(\alpha + \beta)$$

$$y = l_1 \sin \alpha + l_2 \sin(\alpha + \beta) \tag{3.4}$$

$$\theta = \alpha + \beta.$$

The above equations represent *the forward kinematic equations* of the robot—as they relate its control parameters α and β (also known as joint configuration) to the pose of its end-effector in the local coordinate system spanned by x and y, with the origin at the robot's base. Note that both α and β shown in figure 3.1 are positive: both links rotate around the z-axis. Using the right-hand rule, the direction of positive angles is defined to be counter clockwise.

The *configuration space* of the robot—that is, the set of angles each actuator can be set to—is given by $0 < \alpha < \pi$ as it is not supposed to run into the table and $-\pi < \beta < \pi$. The configuration space is defined with respect to the robot's joints and allows us to use the forward kinematics equations to calculate the *workspace* of the robot (i.e., the physical space it can move to). This terminology will be identical for mobile robots. (An example configuration and workspace for both a manipulator and a mobile robot is discussed later.)

We can now write down a transformation that includes a rotation around the z-axis:

$$f(q) = \begin{bmatrix} c_{\alpha\beta} & -s_{\alpha\beta} & 0 & c_{\alpha\beta}l_2 + c_\alpha l_1 \\ s_{\alpha\beta} & c_{\alpha\beta} & 0 & s_{\alpha\beta}l_2 + s_\alpha l_1 \\ 0 & 0 & 1 & 0 \\ 0 & 0 & 0 & 1 \end{bmatrix}. \tag{3.5}$$

The notation $s_{\alpha\beta}$ and $c_{\alpha\beta}$ are shorthand for $sin(\alpha + \beta)$ and $cos(\alpha + \beta)$, respectively. This transformation now allows us to transform from the robot's base to the robot's end-effector configuration $r = [x, y, \theta]^T$ as a function of the joint configuration $q = [\alpha, \beta]^T$. This transformation will be helpful if we want to calculate suitable joint angles in order to reach a certain pose (i.e., inverse kinematics) or if we want to convert measurements taken relative to the end-effector back into the base's coordinate system (e.g., when we have sensors mounted on the end-effector whose output needs to be mapped back to the world reference frame).

3.1.2 The Denavit-Hartenberg Notation

So far, we have considered the forward kinematics of a simple arm and derived relationships between actuator parameters and end-effector positions using basic trigonometry. In the case of multilink arms (the vast majority of robot manipulators in existence), the approach detailed in section 3.1.1 is difficult to scale, so alternative solutions are needed. Interestingly, we can think of the forward kinematics as a chain of homogeneous transformations with respect to a coordinate system mounted at the base of a manipulator (or a fixed position in the room). Deriving these transformations can be confusing and can be facilitated by following a "recipe" such as the one conceived by Hartenberg and Denavit (1955) and Craig (2009). The so-called Denavit-Hartenberg (DH) representation has since evolved as a de facto standard.

A manipulator consists of a series of typically rigid links that are connected by joints. In the vast majority of cases, a joint can either be revolute (i.e., change its angle/orientation) or prismatic (i.e., change its length). Knowing the robot's kinematic properties (e.g., the

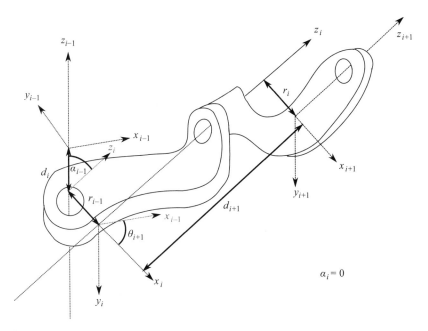

Figure 3.2
Example of selected Denavit-Hartenberg parameters for three sequential revolute joints. The z-axes of joint i and $i+1$ are parallel, which results in $\alpha_i = 0$.

length of all rigid links, similar to l_1 and l_2 in figure 3.1), the pose of its end-effector is fully described by its joint configuration (angle for revolute joints, length for prismatic joints).

In order to use the DH convention, we first need to define a coordinate system at each joint. With reference to figure 3.2, we choose the z-axis to be the axis of rotation for a revolute joint and the axis of translation for a prismatic joint. We can now find the common normal between the z-axes of two subsequent joints (i.e., a line that is orthogonal to each z-axis and intersects both). While the direction of the x-axis at the base can be chosen arbitrarily, subsequent x-axes are chosen such that they lie on the common normal shared between two joints. Whereas the direction of the z-axis is given by the positive direction of rotation (right-hand rule), the x-axis points away from the previous joint. This allows defining the y-axis using the right-hand rule. Note that these rules, in particular the requirement that x-axes lie along the common normal, might result in coordinate systems with their origins outside the joint. This is not problematic because the kinematics of a manipulator is a mathematical representation that need only represent the geometric and kinematic properties of the robot, and it does not need to bear any physical correspondence to the system. The transformation between two joints is then fully described by the following four parameters:

1. The length r (sometimes, a is used) of the common normal between the z-axes of two joints i and $i-1$ (link length).

2. The angle α between the z-axes of the two joints with respect to the common normal (link twist)—that is, the angle between the old and the new z-axis, measured about the common normal.

3. The distance d between the joint axes (link offset)—that is, the offset along the previous z-axis to the common normal.

4. The rotation θ around the common axis along which the link offset is measured (joint angle)—that is, the angle from the old x-axis to the new x-axis, about the previous z-axis.

Two of the above DH parameters describe the link between the joints, and the other two describe the link's connection to a neighboring link. Depending on the link/joint type, these numbers are fixed by the specific mechanical instance of the robot or can be controlled. For example, in a revolute, joint θ is the varying joint angle while all other quantities are fixed. Similarly, for a prismatic joint, d is the joint variable. An example of two revolute joints is shown in figure 3.2.

The final coordinate transform from one link $(i-1)$ to another (i) can now be constructed by concatenating the aforementioned four steps, which are nothing but a series of rotations and translations, one for each DH parameter:

$$_{n-1}^{n}T = T_z'(d_n)R_z'(\theta_n)T_x(r_n)R_x(\alpha_n) \tag{3.6}$$

with

$$T_z'(d_n) = \begin{bmatrix} 1 & 0 & 0 & 0 \\ 0 & 1 & 0 & 0 \\ 0 & 0 & 1 & d_n \\ \hline 0 & 0 & 0 & 1 \end{bmatrix} \quad R_z'(\theta_n) = \begin{bmatrix} \cos\theta_n & -\sin\theta_n & 0 & 0 \\ \sin\theta_n & \cos\theta_n & 0 & 0 \\ 0 & 0 & 1 & 0 \\ \hline 0 & 0 & 0 & 1 \end{bmatrix} \tag{3.7}$$

and

$$T_x(r_n) = \begin{bmatrix} 1 & 0 & 0 & r_n \\ 0 & 1 & 0 & 0 \\ 0 & 0 & 1 & 0 \\ \hline 0 & 0 & 0 & 1 \end{bmatrix} \quad R_x(\alpha_n) = \begin{bmatrix} 1 & 0 & 0 & 0 \\ 0 & \cos\alpha_n & -\sin\alpha_n & 0 \\ 0 & \sin\alpha_n & \cos\alpha_n & 0 \\ \hline 0 & 0 & 0 & 1 \end{bmatrix}. \tag{3.8}$$

These are a translation of d_n along the previous z-axis ($T_z'(d_n)$), a rotation of θ_n around the previous z-axis ($R_z'(\theta_n)$), a translation of r_n along the new x-axis ($T_x(r_n)$), and a rotation of α_n around the new x-axis ($R_x(\alpha_n)$). By replacing each element in equation 3.6, the following matrix is created:

$$_{n-1}^{n}T = \begin{bmatrix} \cos\theta_n & -\sin\theta_n\cos\alpha_n & \sin\theta_n\sin\alpha_n & r_n\cos\theta_n \\ \sin\theta_n & \cos\theta_n\cos\alpha_n & -\cos\theta_n\sin\alpha_n & r_n\sin\theta_n \\ 0 & \sin\alpha_n & \cos\alpha_n & d_n \\ \hline 0 & 0 & 0 & 1 \end{bmatrix}$$

$$= \left[\begin{array}{ccc|c} & R & & t \\ \hline 0 & 0 & 0 & 1 \end{array} \right]. \tag{3.9}$$

where R is the 3×3 rotation matrix and t is the 3×1 translation vector.

Like for any homogeneous transform, the inverse ${}^{n}_{n-1}T^{-1}n$ is given by

$$ {}^{n-1}_{n}T = \left[\begin{array}{ccc|c} & R^{-1} & & -R^{-1}T \\ \hline 0 & 0 & 0 & 1 \end{array} \right] \tag{3.10}$$

with the inverse of R simply being its transpose, $R^{-1} = R^T$.

Similar to the concatenation of transformations detailed in section 2.4.3, ${}^{n}_{n-1}T$ in equation 3.6 can be concatenated with the other transformation matrices relative to the remaining links in order to compute the full kinematics of the robot arm from the base reference frame up to the end-effector.

3.2   Inverse Kinematics

The forward kinematics of a system are computed by means of a transformation matrix from the base of a manipulator (or fixed location, such as the corner of a room) to the end-effector of a manipulator (or a mobile robot). As such, they are an exact description of the pose of the robot, and they fully characterize its kinematic state. Inverse kinematics deal with the opposite problem: finding a joint configuration that results in a desired pose at the end-effector. To achieve this goal, we will need to solve the forward kinematics equations for joint angles as a function of the desired pose. With reference to equation 3.1, inverse kinematics aims to solve the following:

$$q = f^{-1}(r), \qquad f^{-1} : \mathbb{R}^m \rightarrow \mathbb{R}^n, \tag{3.11}$$

with a notation similar to equation 3.1. For a mobile robot, we can do this only for velocities in the local coordinate system and need more sophisticated methods to calculate appropriate trajectories. We will discuss this in depth in section 3.3.

3.2.1   Solvability

Equation (3.11) is the inverted version of equation 3.1 and is heavily nonlinear except for trivial mechanisms. Therefore, it makes sense to briefly think about whether we can solve it at all for specific parameters before trying. Here, the workspace of a robot becomes important. The workspace is the subspace that can be reached by the robot in any configuration. Clearly, there will be no solutions for the inverse kinematic problem outside the workspace of the robot.

A second question to ask is how many solutions we actually expect to exist and what it means to have multiple solutions *geometrically*. Multiple solutions to achieve a desired pose correspond to multiple ways in which a robot can reach a target (i.e., joint configurations). For example a three-link arm that wants to reach a point that can be reached without fully extending all links (which would have only a single solution) can do this by either folding its links in a concave or a convex fashion. Reasoning about how many solutions will exist for a given mechanism and desired pose quickly becomes nonintuitive. For example, a 6-DoF arm can reach certain points with up to 16 different configurations!

3.2.2 Inverse Kinematics of a Simple Manipulator Arm

We will now look at the inverse kinematics of the 2−link arm that we introduced in figure 3.1. We need to solve the equations determining the robot's forward kinematics by solving for α and β. This is tricky, however, as we have to deal with more complicated trigonometric expressions than the forward kinematics case.

To build an intuition, assume there to be only one link, l_1. Solving (3.2) for α yields

$$\alpha = \pm \cos^{-1} \frac{x_1}{l_1}, \tag{3.12}$$

as cosine is symmetric for positive and negative values. Indeed, for any possible position on the x-axis ranging from $-l_1$ to l_1, there exist two solutions: the first one with the arm above the table and the other one with the arm below it. At the extremes of the workspace, both solutions are the same.

Solving (3.4) for α and β adds two additional solutions that are cumbersome to reproduce here, involving terms of x and y to the sixth power; it is left as an exercise to the reader, for example, using an online symbolic solver.

What will drastically simplify this problem is to not only specify the desired position but also the orientation θ of the end-effector. In this case, a desired pose can be specified in the following form:

$$\begin{bmatrix} \cos\theta & -\sin\theta & 0 & x \\ \sin\theta & \cos\theta & 0 & y \\ 0 & 0 & 1 & 0 \\ 0 & 0 & 0 & 1 \end{bmatrix}. \tag{3.13}$$

A solution can now be found by simply equating the individual entries of the transformation (3.5) with those of the desired pose. Specifically, we can observe

$$\cos\theta = \cos(\alpha + \beta) \tag{3.14}$$

$$x = c_{\alpha\beta} l_2 + c_\alpha l_1$$

$$y = s_{\alpha\beta} l_2 + s_\alpha l_1.$$

These equations can be reduced to

$$\theta = \alpha + \beta$$

$$c_\alpha = \frac{c_{\alpha\beta} l_2 - x}{l_1} = \frac{c_\theta l_2 - x}{l_1} \tag{3.15}$$

$$s_\alpha = \frac{s_{\alpha\beta} l_2 - y}{l_1} = \frac{s_\theta l_2 - y}{l_1}.$$

Providing the orientation of the robot in addition to the desired position therefore allows solving for α and β just as a function of x, y, and θ.

The main issue with the geometric approach detailed above is that it does not scale easily with an increase of DoF at the joints, and it quickly becomes unwieldy with more dimensions. For higher-DoF platforms, we can calculate a *numerical solution* using an approach that we will later see is very similar to path planning in mobile robotics. To this end, we will take an optimization-based approach: First, we calculate a measure of error between the current solution and the desired one, and then we change the joint configuration in a way that minimizes this error. In our example, the measure of error is the Euclidean distance between the current end-effector pose (given by the forward kinematics equations in section 3.1.1) and the desired solution $[x, y]$ in configuration space (i.e., assuming $l_1 = l_2 = 1$ for simplicity):

$$f_{x,y}(\alpha, \beta) = \sqrt{\left(s_{\alpha\beta} + s_\alpha - y\right)^2 + \left(c_{\alpha\beta} + c_\alpha - x\right)^2}. \tag{3.16}$$

Here, the first two terms in parentheses are given by the forward kinematics of the robot, whereas the third term in the parentheses is the desired y and x position, respectively. Equation (3.16) can be plotted as a 3D function of α and β, our joint-space variables. As shown in figure 3.3, this function has a minima, in this case zero, for values of α and β that bring the manipulator to $(1, 1)$. These values are $\left(\alpha \to 0, b \to -\frac{\pi}{2}\right)$ and $\left(\alpha \to -\frac{\pi}{2}, b \to \frac{\pi}{2}\right)$.

You can now think about inverse kinematics as a path-finding problem from anywhere in the configuration space to the nearest minima. A more formalized approach will be discussed in section 3.4.2. How to find the shortest path in space—that is, finding the shortest route for a robot to get from A to B—is one of the subjects covered within chapter 13.

3.3   Differential Kinematics

The two-link arm in figure 3.1 involved only two free parameters but was already pretty complex to solve analytically if the end-effector pose was not specified. One can imagine that things become very hard with more degrees of freedom or more complex geometries (mechanisms in which some axes intersect are simpler to solve than others, for example). It is worth noting that, so far, we have analyzed the geometry of motion of a robot at its simplest level of abstraction—that is, in the space of positions. This abstraction becomes useless as soon as the order of motions matters. For example, in a differential-wheel robot,

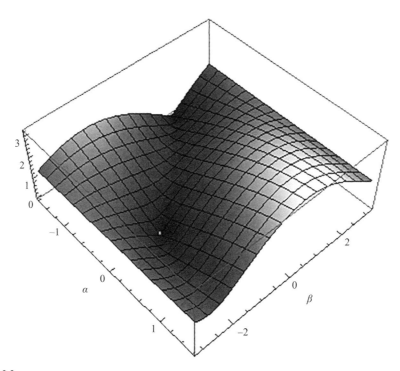

Figure 3.3
Distance to $(x=1, y=1)$ over the configuration space of a two-arm manipulator. Minima corresponds to exact inverse kinematic solutions.

turning the left wheel first and then the right wheel will lead to a very different position than turning the right wheel first and then the left wheel. This is not the case in a robotic arm with two links, which will arrive at the same position no matter which joint is moved first.

In order to include a notion of temporal evolution of the robot configuration, it is convenient to shift toward a slightly more complex abstraction; that is the space of generalized velocities. This modeling is called *differential kinematics* because velocities are the time derivative (i.e., the differential) of positions. As before, with "generalized velocities" we mean any velocity-equivalent quantity needed to describe the element, as detailed below.

3.3.1 Forward Differential Kinematics

Forward differential kinematics deals with the problem of computing an expression that relates the generalized velocities at the joints (i.e., the "speed" of our motors) to the generalized velocity of the robot's end-effector. In all, it is the corresponding differential version of equation 3.1. Let the translational speed of a robot be given by

$$v = \begin{bmatrix} \dot{x} \\ \dot{y} \\ \dot{z} \end{bmatrix}. \tag{3.17}$$

As the robot can potentially not only translate but also rotate, we also need to specify its angular velocity. Let these velocities be given as a vector ω:

$$\omega = \begin{bmatrix} \omega_x \\ \omega_y \\ \omega_z \end{bmatrix}.$$ (3.18)

We can now write translational and rotational velocities in a 6×1 generalized velocity vector as $v = [v\ \omega]^T$. This notation is also called *velocity twist*. Let the generalized configuration in joint space (i.e., joint angles/positions) be $q = [q_1, \ldots, q_n]^T$; therefore, we can define the set of joint speeds as $\dot{q} = [\dot{q}_1, \ldots, \dot{q}_n]^T$. We now want to compute the differential kinematics version of equation 3.1 and in this case relate joint velocities $\dot{q}$ with end-effector velocities $[v\ \omega]^T$. A simple derivation of equation 3.1 with respect to time gives

$$v = [v \quad \omega]^T = J(q) \cdot [\dot{q}_1, \ldots, \dot{q}_n]^T = J(q) \cdot \dot{q},$$ (3.19)

which is our forward differential kinematics equation. $J(q)$ is known as the *Jacobian matrix*; it is a function of the joint configuration q and contains all the partial derivatives of f, relating every joint angle to every velocity. In practice, J looks like the following:

$$v = \begin{bmatrix} v \\ \omega \end{bmatrix} = \begin{bmatrix} \dot{x} \\ \dot{y} \\ \dot{z} \\ \omega_x \\ \omega_y \\ \omega_z \end{bmatrix} = \begin{bmatrix} \frac{\partial x}{\partial q_1} & \frac{\partial x}{\partial q_2} & \cdots & \frac{\partial x}{\partial q_n} \\ \frac{\partial y}{\partial q_1} & \frac{\partial y}{\partial q_2} & \cdots & \frac{\partial y}{\partial q_n} \\ \vdots & \vdots & \vdots & \vdots \\ \frac{\partial \omega_z}{\partial q_1} & \frac{\partial \omega_z}{\partial q_2} & \cdots & \frac{\partial \omega_z}{\partial q_n} \end{bmatrix} \begin{bmatrix} \dot{q}_1 \\ \vdots \\ \dot{q}_n \end{bmatrix} = J(q) \cdot \dot{q}.$$ (3.20)

This notation is important because it tells us how small changes in joint space will affect the end-effector's position in Cartesian space. It may be helpful to think of each column of this matrix as telling us something about how each component of velocity changes when the configuration (i.e., angle) of a particular joint changes, or how each row of the matrix shows how movement in each joint affects a particular component of velocity. The forward kinematics of a mechanism and its analytical derivative can always be calculated, which allows us to calculate numerical values for the entries of matrix J for every possible joint angle/position.

3.3.2   Forward Kinematics of a Differential-Wheel Robot

After abstractly considering differential kinematics in the previous section, we want to now study a mechanism for which general nondifferential kinematic models do not exist. Recall that the pose of a robotic manipulator is uniquely defined by its joint angles, which can be estimated using encoders. However, this is not the case for a mobile robot. Here, the encoder values simply refer to wheel orientations, which need to be integrated over time in order to assess the robot's position with respect to the world's frame of reference. As we will

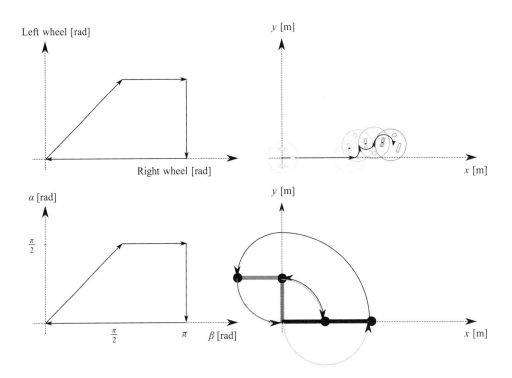

Figure 3.4
Configuration or joint space (left) and workspace or operational space (right) for a non-holonomic mobile robot
(top) and a holonomic manipulator (bottom). Closed trajectories in configuration space result in closed trajectories
in the workspace if the robot's kinematics are holonomic.

later see, this is a source of great uncertainty. What complicates matters is that for so-called
non-holonomic systems, it is not sufficient to simply measure the distance that each wheel
traveled; we must also measure *when* each movement was executed.

A system is non-holonomic when closed trajectories in its configuration space may not
return it to its original state. A robot arm is holonomic because each joint position corresponds to a unique position in space (figure 3.4, bottom). A generic joint-space trajectory
that comes back to the starting point will position the robot's end-effector at the exact same
position in operational space. A train on a track is holonomic, too: Moving the wheels
backward by the same amount they have been moving forward brings the train to the exact
same position in space. Conversely, a car and a differential-wheel robot are non-holonomic
vehicles (figure 3.4, top). Performing a straight line and then a right turn leads to the same
amount of wheel rotation as doing a right turn first and then going in a straight line; however,
getting the robot to its initial position requires not only rewinding both wheels by the same
amount but also getting their relative speeds right. The configuration and corresponding
workspace trajectories for a non-holonomic and a holonomic robot are shown in figure 3.4.

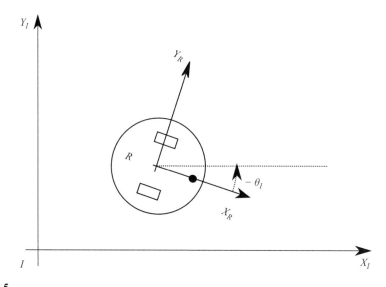

Figure 3.5
Mobile robot with local coordinate system {R} and world frame {I}. The arrows indicate the positive direction of position and orientation vectors.

Here, a robot first moves on a straight line, meaning both wheels turn an equal amount. Then, the left wheel remains fixed and only the right wheel turns forward. Then, the right wheel remains fixed and the left wheel turns backward. Finally, the right wheel turns backward, arriving at the initial encoder values (zero). Yet, the robot does not return to its origin. Performing a similar trajectory in the configuration space of a two-link manipulator makes the robot return to its initial position.

It should be clear by now that for a mobile robot, not only does traveled distance per wheel matter, but so does the *speed* of each wheel as a function of time. Interestingly, this information was not required to uniquely determine the pose of a manipulating arm. We will establish a world coordinate system {I}—which is known as the inertial frame by convention (see figure 3.5). We establish a coordinate system {R} on the robot and express the robot's speed $^R\dot{\xi}$ as a vector $^R\dot{\xi} = [^R\dot{x}, ^R\dot{y}, ^R\dot{\theta}]^T$. Here $^R\dot{x}$ and $^R\dot{y}$ correspond to the speed along the x and y directions in {R}, whereas $^R\dot{\theta}$ corresponds to the rotation velocity around the z-axis that you can imagine to be sticking out of the ground. We denote speeds with dots over the variable name, as speed is simply the derivative of distance. Now, let's think about the robot's position in {R}. It is always zero because the coordinate system is fixed on the robot itself. Therefore, velocities are the only interesting quantities in this coordinate system, and we need to understand how velocities in {R} map to positions in {I}, which we denote by $^I\xi = [^Ix, ^Iy, ^I\theta]^T$. These coordinate systems are shown in figure 3.5.

Notice that the positioning of the coordinate frames and their orientations are arbitrary, meaning it is a choice that we can make. Here, we choose to place the coordinate system in the center of the robot's axle and align $^R x$ with its default driving direction. In order to calculate the robot's position in the inertial frame, we need to first find out how speed in the robot coordinate frame maps to speed in the inertial frame. This can be done again by employing trigonometry. There is only one complication: A movement into the robot's x-axis might lead to movement along both the x-axis and the y-axis of the world coordinate frame I. By looking at figure 3.5, we can derive the following components to $\dot{x}_I$. First,

$$\dot{x}_{I,x} = cos(\theta)\dot{x}_R. \tag{3.21}$$

There is also a component of motion coming from $\dot{y}_R$ (ignoring the kinematic constraints for now, which we will get to below). For negative θ, as in figure 3.5, a move along y_R would let the robot move into positive X_I direction. The projection from $\dot{y}_R$ is therefore given by

$$\dot{x}_{I,y} = -sin(\theta)\dot{y}_R. \tag{3.22}$$

We can now write

$$\dot{x}_I = cos(\theta)\dot{x}_R - sin(\theta)\dot{y}_R. \tag{3.23}$$

Similar reasoning leads to

$$\dot{y}_I = sin(\theta)\dot{x}_R + cos(\theta)\dot{y}_R \tag{3.24}$$

and

$$\dot{\theta}_I = \dot{\theta}_R, \tag{3.25}$$

which is the case because both the robot's coordinate system and the world coordinate system share the same z-axis in this example. We can now conveniently write

$$\dot{\xi}_I = {}^I_R T(\theta)\dot{\xi}_R \tag{3.26}$$

with ${}^I_R T(\theta)$ being a rotation around the z-axis:

$$
{}^I_R T(\theta) = \begin{bmatrix} cos(\theta) & -sin(\theta) & 0 \\ sin(\theta) & cos(\theta) & 0 \\ 0 & 0 & 1 \end{bmatrix}. \tag{3.27}
$$

Maybe unsurprisingly, this is simply the well-known equation for a generic rotation of θ around the z-axis, which applies to both velocity vectors as well as poses.

We are now left with the problem of how to calculate the speed $\dot{\xi}_R$ in robot coordinates. For this, we make use of the *kinematic constraints* of the robotic wheels. For a standard wheel in an ideal-case scenario, the kinematic constraints are that every rotation of the wheel leads to strictly forward or backward motion and does not allow sideways motion or sliding. We can therefore calculate the forward speed of a wheel $\dot{x}$ using its rotational speed $\dot{\phi}$ (assuming the encoder value/angle is expressed as ϕ) and radius r by

$$\dot{x} = \dot{\phi}r. \tag{3.28}$$

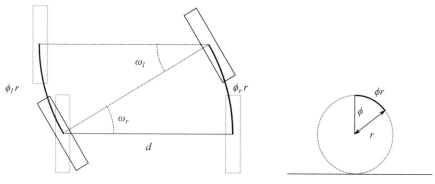

Figure 3.6
Left: Differential-wheel robot pivoting around its left wheel first and its right wheel next. For infinitesimal motion, it is possible to decouple the left and right wheel to simplify computation of the forward kinematics. Right: A wheel with radius r moves by ϕr when rotated by ϕ degrees.

This becomes apparent when considering that the circumference of a wheel with radius r is $2\pi r$. The distance a wheel rolls when turned by the angle ϕ (in radians) is therefore $x = \phi r$ (see figure 3.6, right). Taking the derivative of this expression on both sides leads to the above expression.

How each of the two wheels in our example contributes to the speed of the robot's center—where its coordinate system is anchored—requires the following trick: We calculate the contribution of each individual wheel while assuming all other wheels remaining unactuated (see figure 3.6, left). In this example, the left wheel will move a distance of $r\phi_l$ and the right wheel will move a distance of $r\phi_r$ which in the space of velocities will become $r\dot{\phi}_l$ and $r\dot{\phi}_r$, respectively. Then, the distance traveled by the center point is exactly half of that traveled by each individual wheel (figure 3.6). We can therefore write

$$\dot{x}_R = \frac{1}{2}\left(r\dot{\phi}_l + r\dot{\phi}_r\right) = \frac{r\dot{\phi}_l}{2} + \frac{r\dot{\phi}_r}{2}, \tag{3.29}$$

given the speeds $\dot{\phi}_l$ and $\dot{\phi}_r$ of the left and the right wheel, respectively.

> Think about how the robot's speed along its y-axis is affected by the wheel speed given the coordinate system in figure 3.6. Think about the kinematic constraints that the standard wheels impose.

It may be unintuitive at first, but the speed of the robot along its y-axis is always zero. This is because the constraints of the standard wheel tell us that the robot can never slide. We are now left with calculating the rotation of the robot around its z-axis. This rotation can be seen when imagining the robot's wheels spinning in opposite directions. In this case, the robot does not move forward but rather spins in place. We will again consider each wheel independently. Assuming the left wheel to be non-actuated, spinning the right wheel forward will lead to counter clockwise rotation. Given an axle diameter (distance between the robot's wheels) d, we can now write

$$\omega_r d = \phi_r r \tag{3.30}$$

with ω_r the angle of rotation around the left wheel (figure 3.6, left). Taking the derivative on both sides yields speeds and we can write

$$\dot{\omega}_r = \frac{\dot{\phi}_r r}{d}. \tag{3.31}$$

Adding the rotation speeds (with the one around the right wheel being negative based on the right-hand grip rule) leads to:

$$\dot{\theta} = \frac{\dot{\phi}_r r}{d} - \frac{\dot{\phi}_l r}{d}. \tag{3.32}$$

Putting it all together, we can finally write:

$$\begin{bmatrix} \dot{x}_I \\ \dot{y}_I \\ \dot{\theta} \end{bmatrix} = \begin{bmatrix} cos(\theta) & -sin(\theta) & 0 \\ sin(\theta) & cos(\theta) & 0 \\ 0 & 0 & 1 \end{bmatrix} \begin{bmatrix} \frac{r\dot{\phi}_l}{2} + \frac{r\dot{\phi}_r}{2} \\ 0 \\ \frac{\dot{\phi}_r r}{d} - \frac{\dot{\phi}_l r}{d} \end{bmatrix}. \tag{3.33}$$

The interested reader might want to compare this form with equation 3.20, the general Jacobian form of differential kinematics. For this, we ignore the rotation matrix in equation 3.33 and rewrite its second term in matrix notation,

$$\begin{bmatrix} \dot{x}_R \\ \dot{y}_R \\ \dot{\theta} \end{bmatrix} = \begin{bmatrix} \frac{r}{2} & \frac{r}{2} \\ 0 & 0 \\ \frac{r}{d} & -\frac{r}{d} \end{bmatrix} \begin{bmatrix} \dot{\phi}_l \\ \dot{\phi}_r \end{bmatrix} = \begin{bmatrix} \frac{\partial x_R}{\partial \phi_l} & \frac{\partial x_R}{\partial \phi_r} \\ \frac{\partial y_R}{\partial \phi_l} & \frac{\partial y_R}{\partial \phi_r} \\ \frac{\partial \theta}{\partial \phi_l} & \frac{\partial \theta}{\partial \phi_l} \end{bmatrix} \begin{bmatrix} \dot{\phi}_l \\ \dot{\phi}_r \end{bmatrix}, \tag{3.34}$$

then with $X_R = \left(\frac{r\dot{\phi}_l}{2} + \frac{r\dot{\phi}_r}{2} \right) t$ and similar expressions for θ, we observe the validity of the Jacobian approach.

From forward kinematics to odometry

Equation (3.33) only provides us with the relationship between the robot's wheel speed and its speed in the inertial frame. Calculating its actual pose in the inertial frame is known as *odometry*. Technically, it requires integrating equation 3.33 from 0 to the current time T. Since this is not possible but for very special cases, one can approximate the robot's pose by summing up speeds over discrete time intervals, or more precisely,

$$\begin{bmatrix} x_I(T) \\ y_I(T) \\ \theta(T) \end{bmatrix} = \int_0^T \begin{bmatrix} \dot{x}_I(t) \\ \dot{y}_I(t) \\ \dot{\theta}(t) \end{bmatrix} dt \approx \sum_{k=0}^{k=T} \begin{bmatrix} \Delta x_I(k) \\ \Delta y_I(k) \\ \Delta \theta(k) \end{bmatrix} \Delta t, \tag{3.35}$$

which can be calculated incrementally as

$$x_I(k+1) = x_I(k) + \Delta x(k) \Delta t \tag{3.36}$$

using $\Delta x(k) \approx \dot{x}_I(t)$ and similar expressions for y_I and θ. Note that equation 3.36 is just an approximation. The larger Δt becomes, the more inaccurate this approximation becomes, since the robot's speed might change during the interval.

Don't let the notion of an integral worry you! Because a robot's computers are fundamentally discrete, integrals usually turn into sums, which are nothing more complex than for-loops.

3.3.3 Forward Kinematics of Carlike Steering

Differential-wheel drives are very popular in mobile robotics because they are very easy to build, maintain, and control. Although not holonomic, a differential drive can approximate the function of a fully holonomic robot by first rotating in place to achieve a desired heading and then driving straight. The primary drawbacks of a differential drive are its reliance on a caster wheel, which performs poorly at high speeds, and difficulties in driving straight lines as it requires both motors to drive at the exact same speed.

These drawbacks are mitigated by carlike mechanisms, which are driven by a single motor and can steer their front wheels. This mechanism is known as "Ackermann steering." Ackermann steering should not be confused with "turntable" steering where the front wheels are fixed on an axis with central pivot point. Instead, in Ackermann steering, each wheel has its own pivot point, and the system is constrained in such a way that all wheels of the car trace circles with a common center point, avoiding skid. As the Ackermann mechanism lets all wheels drive on circles with a common center point, its kinematics can be approximated by those of a tricycle with rear-wheel drive or, even simpler, by a bicycle. This is shown in figure 3.7.

Consider a car with the shape of a box with length L between its front and rear axis. Let the center point of the common circle described by all wheels be distance R from the car's longitudinal center line. Then, the steering angle ϕ is given by

$$\tan \phi = \frac{L}{R}. \tag{3.37}$$

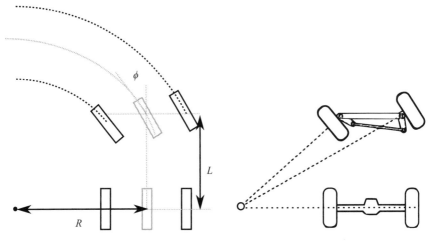

Figure 3.7
Left: Kinematics of carlike steering and the equivalent bicycle model. Right: Mechanism of an Ackermann vehicle.

The angles of the left and the right wheel, ϕ_l and ϕ_r can be calculated using the fact that all wheels of the car rotate around circles with a common center point. With the distance between the two front wheels l, we can write:

$$\frac{L}{R - l/2} = \tan(\phi_r)$$

$$\frac{L}{R + l/2} = \tan(\phi_l). \tag{3.38}$$

This is important, because it allows us to calculate the resulting wheel angles resulting from a specific angle ϕ and test whether they are within the constraints of the actual vehicle.

Assuming the wheel speed to be $\dot{\omega}$ and the wheel radius r, we can calculate the speeds in the robot's coordinate frame as

$$\dot{x}_r = \dot{\omega} r$$

$$\dot{y}_r = 0 \tag{3.39}$$

$$\dot{\theta}_r = \frac{\dot{\omega} r \tan \phi}{L}$$

using (3.37) to calculate the circle radius R.

3.4   Inverse Differential Kinematics

It would now be desirable to just invert J in equation 3.20 in order to calculate the necessary joint speeds for every desired end-effector speed—a problem known as *inverse differential kinematics*. Unfortunately, J is only invertible if the matrix is quadratic (i.e., the number of degrees of freedom in joint space n equals the number of degrees of freedom in task space m) and full rank. In the example detailed in section 3.2.2, the velocity wrench $[v\ \omega]^T$ is six-dimensional, which means that n should be equal to 6. Therefore, inversion of J is only possible if the robot under consideration is equipped with exactly six actuators/joints. If this is not the case, we can use the pseudo-inverse computation:

$$J^+ = \frac{J^T}{JJ^T} = J^T(JJ^T)^{-1}. \tag{3.40}$$

As you can see, J^T cancels from the equation leaving $1/J$, while being applicable to non-quadratic matrices. We can now write a simple feedback controller that drives our error e, defined as the difference between desired and actual position, to zero:

$$\Delta q = -J^+ e. \tag{3.41}$$

That is, we will move each joint a tiny bit in the direction that minimizes our error e. It can be easily seen that the joint speeds will only be zero if e has become zero.

This solution might or might not be numerically stable, depending on the current joint values. If the inverse of J is mathematically not feasible, we speak of a *singularity* of the mechanism. One case where this can happen is when two joint axes line up, therefore effectively removing a degree of freedom from the mechanism, or when the robot is at the boundary of its workspace. As it happens, very often in robotics, the concept of singularity has both a strong mathematical justification (the joint configuration is such that the Jacobian is not full rank any more) and a direct physical consequence: Singularity configurations are to be avoided as no solution for the inverse differential kinematics problem exists and the robot might become unsafe to operate. In a singularity, the solution to J^+ "explodes" and joint speeds go to infinity. In order to work around this, we can introduce damping to the controller.

In this case, we do not only minimize the error but also the joint velocities. The minimization problem then becomes:

$$\|J\Delta q - e\| + \lambda^2 \|\Delta q\|^2, \tag{3.42}$$

where λ is a constant. One can show that the resulting controller that achieves this has the form

$$\Delta q = (J^T J + \lambda^2 I)^{-1} J^+ e. \tag{3.43}$$

This is known as the *damped least-squares* method. Problems that can arise when taking this approach include the existence of local minima and singularities of the mechanism, which might render solutions suboptimal or infeasible.

3.4.1 Inverse Kinematics of Mobile Robots

There is no unique relationship between the amount of rotation of a robot's individual wheels and its position in space; however, for simple holonomic platforms such as a robot on a track, we can treat the inverse kinematics problem as solving for the velocities within the local robot coordinate frame. Let's first establish how to calculate the necessary speed of the robot's center given a desired speed $\dot{\xi}_I$ in world coordinates. We can transform the expression $\dot{\xi}_I = T(\theta)\dot{\xi}_R$ by multiplying both sides with the inverse of $T(\theta)$,

$$T^{-1}(\theta)\dot{\xi}_I = T^{-1}(\theta)T(\theta)\dot{\xi}_R, \tag{3.44}$$

which leads to $\dot{\xi}_R = T^{-1}(\theta)\dot{\xi}_I$. Here

$$T^{-1} = \begin{bmatrix} cos\theta & sin\theta & 0 \\ -sin\theta & cos\theta & 0 \\ 0 & 0 & 1 \end{bmatrix}, \tag{3.45}$$

which can be computed by performing the matrix inversion or by deriving the trigonometric relationships from the drawing. Similar to the approach in section 3.2.2, we can now solve equation 3.33 for ϕ_l, ϕ_r:

$$\dot{\phi}_l = (2\dot{x}_R - \dot{\theta}d)/2r \tag{3.46}$$

$$\dot{\phi}_r = (2\dot{x}_R + \dot{\theta}d)/2r,$$

allowing us to calculate the robot's wheel speed as a function of a desired $\dot{x}_R$ and $\dot{\theta}$, which can be calculated using (3.44).

Note that this approach does not allow us to deal with $\dot{y}_R \neq 0$, which might result from a desired speed in the inertial frame. Nonzero values for translation in y-direction are simply ignored by the inverse kinematic solution, and driving toward a specific point either requires feedback control (section 3.4.2) or path planning (chapter 13).

Inverse kinematics of an omnidirectional robot

Omnidirectional robots using "Swedish wheels" or "Meccano wheels" are common in factories and educational settings. A drawing of a Swedish wheel is shown in table 2.1. It consists of an actuated wheel with non-actuated rollers around its circumference that are attached in a 45-degree angle. Similar to the caster wheel, the Swedish wheel has full three degrees of freedom in the plane but can enable omnidirectional motion of a robotic platform without the need to rotate. A typical four-wheel configuration is shown in figure 3.8. Notice the arrangement of the wheels and, in particular, the orientation of the rollers, which is critical for the operation as shown.

When actuated by itself, the wheel will perform a sideways motion that is perpendicular to the main axis of its rollers. When used in pairs, opposite directions of motions cancel

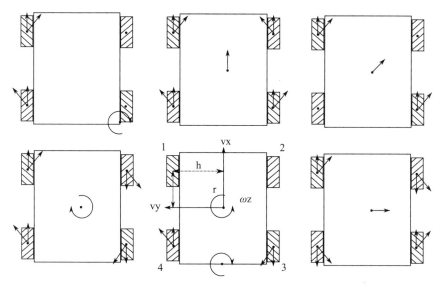

Figure 3.8
Omnidirectional robot using "Swedish wheels" in different configurations. Each wheel has two velocity components: speed perpendicular to the wheel's main axis and speed of the rollers. Arrows on the robot body indicate the resulting direction of motion and rotation.

out, resulting in forward motion (as shown in figure 3.8, top, center) or sideways motion (figure 3.8, bottom, right).

Similar to a differential-wheel platform, each wheel also imparts a rotation on the robot body. As the wheels are mounted off the center axle, each wheel contributes to two angular moments. One is around the horizontal axis with distance h_i to the robot's center, and the other is around the vertical axis with distance r_i to the robot's center (figure 3.8, bottom, center). All combined, the rotation of each wheel will add up to the robot moving with velocities v_x, v_y, and ω_z.

The velocity at each wheel has two components: the velocity of the i-th wheel perpendicular to its main axis $v_{i,m}$, and the velocity of the rollers $v_{i,r}$ that is either $+45$ degrees or -45 degrees to the wheel axis. Note that for the system to work, diagonally opposite wheels need to have the same angle. Let the angle of the roller of wheel i be γ_i. We can now derive, following Maulana, Muslim, and Hendrayawan (2015), an equation that reads

$$v_{i,m} + v_{i,r}\cos(\gamma_i) = v_x - h_i * \omega_z. \tag{3.47}$$

That is, the velocity components perpendicular to the wheel axis are equivalent to the forward velocity of the robot plus the velocity component at the wheel resulting from the robot's angular velocity. Positive angular velocity would result in backward motion, by definition of the robot coordinate system. Similarly, we can write

$$v_{i,r}\sin(\gamma_i) = v_y + r_i * \omega_z. \tag{3.48}$$

Note that there is no lateral component to the main wheel's motion, because lateral motion can only be achieved with the rollers.

Dividing (3.48) by (3.47) and solving for v_i results in

$$v_i = v_x - h_i\omega_z - \frac{v_y + r_i\omega_z}{\tan\gamma_i}. \tag{3.49}$$

With $h_i \in h = \{h, -h, h, -h\}$, $r_i \in r = \{r, r, -r, -r\}$, and $\gamma_i \in \gamma = \{-45\deg, +45\deg, +45\deg, -45\deg\}$ to reflect the different configuration of each wheel, we can derive the following expression for the controllable wheel velocities $v_{i,m}$:

$$v_{1,m} = v_x + v_y + r\omega_z - h\omega_z \tag{3.50}$$

$$v_{2,m} = v_x - v_y - r\omega_z + h\omega_z$$

$$v_{3,m} = v_x - v_y + r\omega_z - h\omega_z$$

$$v_{4,m} = v_x + v_y - r\omega_z + h\omega_z.$$

With $v_{i,m} = R\omega_i$ and R the radius of each Swedish wheel, we can now compute the required wheel velocity for any desired robot velocity v_x, v_y, and ω_z.

3.4.2   Feedback Control for Mobile Robots

Assume the robot's position to be given by x_r, y_r, and θ_r and the desired pose as x_g, y_g, and θ_g—with the subscript g indicating "goal." We can now calculate the error in the desired pose by using

$$\rho = \sqrt{(x_r - x_g)^2 + (y_r - y_g)^2}$$

$$\alpha = \tan^{-1}\frac{y_g - y_r}{x_g - x_r} - \theta_r \tag{3.51}$$

$$\eta = \theta_g - \theta_r,$$

which is illustrated in figure 3.9. These errors can be converted directly into robot speeds using, for example, a simple proportional controller with gains p_1, p_2, and p_3:

$$\dot{x} = p_1\rho \tag{3.52}$$

$$\dot{\theta} = p_2\alpha + p_3\eta, \tag{3.53}$$

which will let the robot drive in a curve until it reaches the desired pose.

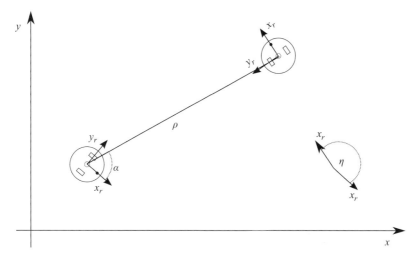

Figure 3.9
Difference in desired and actual pose as a function of distance ρ, bearing α, and heading η.

3.4.3 Under-Actuation and Over-Actuation

As detailed at the beginning of this chapter, kinematics is concerned with analyzing the mapping between our control variables (i.e., the robot's motors represented by the n DoFs in joint space) and their effect on the motion of the robot (our m DoFs in task/configuration space). These two spaces might have different dimensionality, and the relation between these two dimensions greatly affects how we can solve the kinematic problem. It is convenient to analyze these differences by looking at the Jacobian J, since the size of the matrix is $m \times n$; in all, we have three different conditions:

- $n = m \rightarrow$ The robot is *fully actuated*. The Jacobian J is square and full rank, and the forward kinematics equation is directly invertible.

- $n \leq m \rightarrow$ The robot is *under actuated*, and the kinematics problem is *kinematically deficient*. The Jacobian J is "wide," because there are more columns m than rows n, and not invertible anymore; the only way to solve the inverse kinematics problem is through the pseudo-inverse J^+ (and similar/more advanced approaches).

- $n \geq m \rightarrow$ The robot is *over actuated*, and the kinematics problem is *kinematically redundant*. The Jacobian J is "tall," because there are more rows n than columns m, and not invertible anymore; the only way to solve the inverse kinematics problem is through the pseudo-inverse J^+ (and similar/more advanced approaches). In this scenario, it is useful to determine the redundancy coefficient $n - m$ that affects the space of solutions of the inverse kinematics problem.

Over- and under-actuation are important design considerations to keep in mind when choosing a robot for a particular task. In a *kinematically deficient* scenario, the robot is not capable of full motion in task space because it does not have sufficient degrees of freedom in joint space to "cover" every possible configuration in task space. This does not mean that the robot is useless! It can still perform tasks—just not *every* possible task you might ask it to perform. Conversely, if the problem is *kinematically redundant*, the robot has more joint DoFs available than it needs, and there exist an infinite number of inverse kinematics solutions in non-singular configurations. Contrary to what one may think, redundancy is actually a great feature to have in a robot system because it allows for flexibility and versatility in solving the kinematic problem—that is, it is possible to choose *the best* solution among many, and one that satisfies additional constraints and requirements. A human arm (without considering the hand) is a good example of a kinematically redundant manipulator. A person's arm is equipped with seven DoFs in joint space (three at the shoulder, one at the elbow, and three at the wrist), whereas the task space is of dimension six (i.e., the three positions and three orientations of the wrist). This additional degree of mobility allows humans to reach for objects in multiple configurations, choose motions that are energy efficient, and avoid obstacles.

Take-Home Lessons

• Forward kinematics deals with transforming from a world coordinate system to a coordinate system on a robot. Such a transform is a combination of a (3×1) translation vector and a (3×3) rotation matrix that consists of the unit vectors of the robot coordinate system. Both translation and rotation can be combined into a 4×4 homogeneous transform matrix.

• Forward and inverse kinematics of a mobile robot are performed with respect to the speed of the robot and not its position.

• To calculate the effect of each wheel on the speed of the robot, you need to consider the contribution of each wheel independently.

• The inverse kinematics of a robot involve solving the forward kinematics equations for the joint angles. Calculating the inverse kinematics analytically becomes quickly infeasible and is impossible for complicated mechanisms.

• A simple numerical solution is provided by taking all partial derivatives of the forward kinematics in order to get an easily invertible expression that relates joint speeds to end-effector speeds.

• The inverse kinematics problem can then be formulated as a feedback control problem, which will move the end-effector toward its desired pose using small steps. Problems with this approach are local minima and singularities of the mechanism, which might render this solution infeasible.

• Redundancy allows a robot to solve a kinematic problem in multiple ways, thus providing better dexterity and versatility in its motion.

Exercises

Coordinate Systems

1. Write out the entries of a rotation matrix $^A_B R$ assuming basis vectors X_A, Y_A, Z_A, and X_B, Y_B, Z_B. Then write out the entries of rotation matrix $^B_A R$.

2. Assume two coordinate systems that are colocated in the same origin but rotated around the z-axis by the angle α. Derive the rotation matrix from one coordinate system to the other, and verify that each entry of this matrix is indeed the scalar product of each basis vector of one coordinate system with every other basis vector in the second coordinate system.

3. Consider two coordinate systems, $\{B\}$ and $\{C\}$, whose orientation is given by the rotation matrix $^C_B R$ and that have distance $^B P$. Provide the homogenous transform $^C_B T$ and its inverse $^B_C T$.

4. Consider the frame $\{B\}$ that is defined with respect to frame $\{A\}$ as $\{B\} = \{^A_B R, ^A P\}$. Provide a homogeneous transform from $\{A\}$ to $\{B\}$.

5. Program a simple application that displays a 2D (or 3D) coordinate system and add the ability to move and turn the coordinate system using your keyboard.

Forward and Inverse Kinematics

1. Consider a differential-wheel robot with a broken motor (i.e., one of the wheels cannot be actuated anymore). Derive the forward kinematics of this platform. Assume the right motor is broken.

2. Consider a tricycle with two independent standard wheels in the rear and the steerable, driven front wheel. Choose a suitable coordinate system and use ϕ as the steering-wheel angle and wheel speed $\dot{\omega}$. Provide forward and inverse kinematics.

3. Program an application that displays a differential-wheel platform and allows you to control forward and rotational speed with your keyboard. Output the robot's pose after every step.

4. Program an application that displays a two-link robotic arm moving in the plane and lets you change both joint angles using your keyboard.

5. Derive the forward kinematics of a two-link robotic arm as well as its Jacobian. Implement its inverse kinematic using the inverse Jacobian technique.

6. Program an application that displays a two-link robotic arm moving in the plane and lets you change the position of its end-effector using your keyboard.

7. Explore the internet for toolkits that allow you to manage forward and inverse kinematics for a robotic arm. What kind of tools do you find, and what kind of input do they require to model the robot's geometry?

8. Download the manual of a commercially available robot arm of your choice. What kind of input does it take from its user? Does it allow you to control its position directly?

9. Use a robot simulator of your choice to access a simulated vehicle. What kind of actuator input can you provide, and what are the sensors that are available? Drive the car using your keyboard and try to estimate its pose by implementing odometry.

10. Can you provide an example of a kinematically deficient and a kinematically redundant robot manipulator?

4 Forces

So far, we have only been concerned with how robots move and the *geometry of motion*. However, moving a robot not only requires a kinematic model of the platform under consideration but also an understanding of the (generalized) forces needed to actuate the robot's motors and those needed for the robot to interact with the environment. While this aspect can be ignored in basic applications of mobile robots and simple manipulation, it becomes critical as soon as robots interact more closely with people or need to engage in more complex manipulation: in these scenarios, *safety* and *model accuracy* are of paramount importance.

In this chapter, we will introduce the reader to these concepts through *statics*, introducing a third abstraction to the problem of analyzing how robots move in space and interact with their surroundings. More specifically, in sections 3.1 and 3.2 we have investigated the *kinematic* problem and operated in the space of *positions*—that is, how to map joint angles with end-effector poses. In section 3.3, we introduced the *differential kinematics* problem and operated in the space of *velocities*—that is, how to map joint velocities with end-effector velocity twists (remember, velocity is derivative of position, hence the name "differential"). Now we will operate in the space of *forces*; however, we will simplify the more general dynamical problem by looking at the robot in static equilibrium—otherwise known as a *static* configuration. A lot can be done by simply looking at the robot in an equilibrium configuration! The fourth and last abstraction, which goes beyond the scope of this book, is called *dynamics* and operates in the space of forces from a non-static perspective; it involves the second derivative of position (i.e., the acceleration), and it can be thought as a generalization of the second law of Newton ($F = ma$). The goals of this chapter are to introduce readers to the following concepts:

- The concept of statics
- The so-called "kineto-statics duality"
- The notion of "manipulability"

Most of the concepts that we discuss here are typically considered in the context of manipulation, as mobile robots generally do not exchange forces with their environment.

Therefore, for simplicity, we will hereinafter refer to robot manipulators equipped with revolute joints unless otherwise specified.

> The analysis of motion of a robot can be thought of as a layered system with multiple levels of abstraction of increasing complexity. The more complex it becomes, the more comprehensive your analysis will be, and the more capability you will be able to squeeze out of the robot! However, it is good practice to start with the simplest layer first (i.e., kinematics) and gradually progress toward a dynamic analysis only if needed.

4.1 Statics

Statics deals with relating (generalized) forces at the robot's joints and generalized forces at the end-effector when the robot is in *static (or mechanical) equilibrium* (i.e., the acceleration of the robot and all its components is zero). If such a condition is met, a robot with n degrees of freedom and an end-effector characterized by m degrees of freedom can be fully described by the following quantities:

- An $(n \times 1)$ vector of generalized forces τ at the joints.
- An $(m \times 1)$ vector of generalized forces F exerted *by the robot end-effector* on the environment—or, more generally, by any part of the robot that may be in direct physical contact with the environment.
- An $(m \times 1)$ vector of forces exerted *by the environment* on the robot end-effector F_e—which, per the principle of action and reaction, are equal and opposite to F: $F_e = -F$.

In this case, *generalized force* means "any force-equivalent quantity needed to describe the element." In the case of joints, it depends on the actuation: Generalized forces at the joints are either forces for prismatic joints (as they impart a translational motion on the joint) or torques for revolute joints (as they impart a rotational motion on the joint); the size of this vector depends on the number of mechanical degrees of freedom (DoF) n. In the case of the end-effector, it depends on the number of DoFs in task space m. If we are operating with a 6-DoF problem, the $m \times 1$ vector of generalized forces will be composed of a linear force component given by the forces on the three axes,

$$f = \begin{bmatrix} f_x \\ f_y \\ f_z \end{bmatrix}, \tag{4.1}$$

and an angular force component (or moment) μ around the three axes, given as

$$\mu = \begin{bmatrix} \mu_x \\ \mu_y \\ \mu_z \end{bmatrix}. \tag{4.2}$$

We can now combine the above elements in a 6×1 vector as $F = [f \ \mu]^T$. This vector of generalized forces is also called a *spatial force* or *wrench*. We now want to compute the statics version of equations (3.19) and (3.20) and relate our $n \times 1$ vector of torques τ with the 6×1 wrench vector F. To find this relationship, let's recall the definition of *power* from physics. Mechanical power W is defined as force times velocity, which can be generalized as generalized forces times generalized velocities: $W = F^T \cdot v$. Now, we know that the forces exchanged at the end-effector come from our source of actuation (i.e., our motors) whose generated power is defined by $W = \tau^T \cdot \dot{q}$. We therefore have that

$$W = \tau^T \cdot \dot{q} = F^T \cdot v. \tag{4.3}$$

We also know the relation between v and $\dot{q}$ from equation (3.19): $v = J(q) \cdot \dot{q}$. Equation (4.3) then becomes

$$\tau^T \cdot \dot{q} = F^T \cdot J(q) \cdot \dot{q}, \tag{4.4}$$

which, with minor rearrangements, turns into the following:

$$\tau = J(q)^T \cdot F. \tag{4.5}$$

This is the final statics equation we were looking for! It can be interpreted as follows: To counteract an external wrench $F_e = -F$ applied on the end-effector by the environment in a static configuration q, the robot needs to apply torques τ at its joints as specified by equation (4.5). Interestingly, this equation clearly shows how statics acts as a middle ground between the "geometry-only" kinematics approach and the more general dynamics problem: Even though we are dealing with forces and torques, their relationship is defined by a geometric relation—that is, the same Jacobian used in section 3.4. In this case, we are using its $n \times m$ transpose:

$$\tau = \begin{bmatrix} \tau_1 \\ \vdots \\ \tau_n \end{bmatrix} = \begin{bmatrix} \frac{\partial x}{\partial q_1} & \frac{\partial y}{\partial q_1} & \cdots & \frac{\partial \omega_z}{\partial q_1} \\ \frac{\partial x}{\partial q_2} & \frac{\partial y}{\partial q_2} & \cdots & \frac{\partial \omega_z}{\partial q_2} \\ \vdots & \vdots & \vdots & \vdots \\ \frac{\partial x}{\partial q_n} & \frac{\partial y}{\partial q_n} & \cdots & \frac{\partial \omega_z}{\partial q_n} \end{bmatrix} \begin{bmatrix} f_x \\ f_y \\ f_z \\ \mu_x \\ \mu_y \\ \mu_z \end{bmatrix} = J(q)^T \cdot F. \tag{4.6}$$

Equation (4.5) is useful on a variety of different problems. The most typical application is *force control*—that is, the robot's motors are actuated so as to apply a specified wrench on the environment. For example, one may want to use a robot for a polishing task in which it needs exert a vertical force of $5N$ on a table. In this case, the desired wrench (assuming our z-axis in Cartesian space is the vertical one and it is pointing upward) would be:

$$F = \begin{bmatrix} 0 \\ 0 \\ -5N \\ 0 \\ 0 \\ 0 \end{bmatrix}. \tag{4.7}$$

4.2   Kineto-Statics Duality

The analogy between equations (3.20) and (4.5) by means of the Jacobian makes it interesting to analyze equation (4.5) similarly to what we did in sections 3.3 and 3.4.3. This analogy is defined as *kineto-statics duality* and helps the novice roboticist to more intuitively correlate these two levels of abstraction. More specifically, singular configurations are as relevant to the statics problem as they are to the differential kinematics one, but they have different physical interpretation. In a singular configuration, both the Jacobian and its transpose lose rank—that is because transposing a matrix does not affect its rank. However, while loss of full rank affects the *inverse* kinematics problem (i.e., its solution "explodes" and joint speeds go to infinity), in this case it is the *direct* statics mapping that is affected by it. In a singular configuration, forces exerted by the robot on the environment go to infinity. This is an additional (and arguably more compelling) reason to avoid singularities at all costs: the robot would move very fast *and* exert strong forces on anything on its path.

4.3   Manipulability

The duality property that exists between differential kinematics (section 3.3) and statics (section 4.1) allows us to further inspect manipulator performance for a given joint configuration.

4.3.1   Manipulability Ellipsoid in Velocity Space

As a first step, we may inspect the capacity of the manipulator to arbitrarily change its end-effector's position and orientation from the current configuration. More specifically, we may ask the following question: What effect does a small increment in joint positions (i.e., a small joint velocity) have on the end-effector pose? Let's consider the set of joint velocities of unit norm defined by the following equation:

$$\dot{q}^T \cdot \dot{q} = 1. \tag{4.8}$$

This equation represents a multidimensional "sphere" in joint space $\mathbb{R}^n$. We know from section 3.3 that this corresponds to a similarly multidimensional shape in operational space $\mathbb{R}^m$, and we know that this correspondence is mediated by equation (3.19) and its inverse. In the generic case of a redundant manipulator, equation (4.8) becomes

$$v^T J(q)^{+T} \cdot J(q)^+ v = 1, \tag{4.9}$$

which, combined with equation (3.40), becomes

$$v^T \left[J(q) \cdot J(q)^T \right]^{-1} v = 1, \tag{4.10}$$

which corresponds to a multidimensional ellipsoid in operational space m—otherwise known as *velocity manipulability ellipsoid*. This ellipsoid provides the following physical interpretations:

• Along the direction of its major axis, the robot can move at large velocities.
• Along the direction of its minor axis, the robot can move at small velocities.
• The volume of the ellipsoid is called *manipulability measure* and is defined as $w(q) = \sqrt{det \left[J(q)J(q)^T \right]}$. It is always positive, and it reaches a maximum when the ellipsoid is close to a sphere and the robot can move isotropically in any direction.
• In a singularity, the ellipsoid "loses a dimension" and one of its axis becomes 0. Concurrently, the manipulability measure $w(q) = 0$—which is why $w(q)$ is used to understand how far a robot is from a singular configuration.

The properties above can easily be verified in the top half of figure 4.1. The closer the robot is to a singular configuration (e.g., the arm fully stretched), the more the two-dimensional ellipsoid converges to a vertical line. At the singularity itself, the minor axis has zero length, signifying that the robot can only move on a vertical direction and not right or left.

4.3.2 Manipulability Ellipsoid in Force Space

By virtue of the kineto-statics duality, similar considerations can be done in force space. In this case, we may want to consider a sphere in the space of joint torques,

$$\tau^T \cdot \tau = 1, \tag{4.11}$$

which, thanks to equation (4.5), is mapped into a *force manipulability ellipsoid*:

$$F^T \left[J(q) \cdot J(q)^T \right] F = 1. \tag{4.12}$$

This ellipsoid characterizes the forces at the end-effector that can be exerted by the robot on the environment in the given joint configuration q. It behaves similar to the velocity

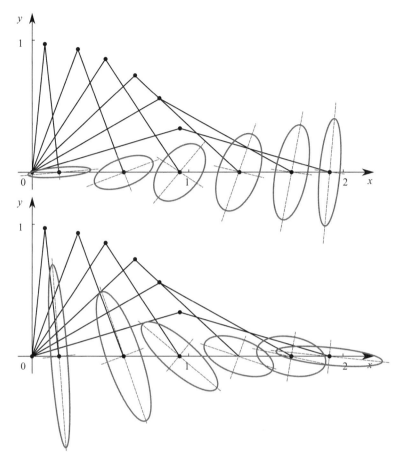

Figure 4.1
Velocity (top) and force (bottom) manipulability ellipsoids for a 2-DoF planar arm ($n = m = 2$). In this simple 2×2 case, the ellipsoids collapse to simple ellipses (whose major and minor axes are drawn by a dotted line).

manipulability ellipsoid, with one important difference: While the principal axes of both ellipsoids are in the same orientation, their magnitude is in inverse proportions. As depicted in figure 4.1 (bottom), the major axis in force space becomes the minor axis in velocity space and vice versa. Therefore, a direction of high velocity manipulability corresponds to a direction of low force manipulability.

4.3.3   Manipulability Considerations

The velocity and force manipulability ellipsoids are useful for a variety of tasks, from identifying a suitable joint configuration to perform a specific task to understanding what it is possible for the robot to do in a specific configuration. Remember that for a kinematically redundant manipulator (see section 3.4.3), it is possible to be in the same task space configuration (e.g., end-effector pose) with multiple joint postures. Therefore, a manipulability analysis may allow the robot designer to choose the configuration that better conforms to additional specifications (e.g., lower exerted force, lower energy consumption, and better legibility of robot motion by humans).

Take-Home Lessons

1. Looking at forces in mechanical equilibrium—that is, when end-effector forces and joint torques cancel each other—allows us to extend control of the robot from poses and velocities to the force domain.

2. The torques required by a robotic arm are related to end-effector forces using the same Jacobian that also defines the robot's differential kinematics—a concept known as the kineto-statics duality.

3. Although a robotic arm might reach a desired pose using multiple different configurations, some configurations are better suited than others; a manipulability analysis helps in characterizing this problem.

Exercises

1. Think about the four layers of abstraction we have just investigated: kinematics, differential kinematics, statics, and dynamics.

a) Can you think of an application for which you would need a dynamic analysis? (Hint: This is generally something really hard.)

b) What can be done by just looking at the static problem instead? (Hint: You are still considering an exchange of forces here.)

c) What can you do with a robot from a purely kinematic perspective? (Hint: This is typically easy.)

2. Why are singular configurations dangerous for the robot and its surroundings? Think about the relationship between forces and velocities.

3. How can you ensure the robot "stays away" from singularities?

4. Program an application that displays the manipulability ellipsoids in force and velocity for a two-link planar arm (similar to figure 4.1). Feel free to integrate this program with the kinematics exercise in chapter 3. How does the manipulability ellipsoid relate to positional increments of the end-effector? What happens in a singularity? (Hint: The easiest singularity to find for robot manipulators is the "stretched out" configuration.)

5. Use a robot simulator of your choice to access a robot manipulator with at least three DoFs in joint space that moves in 3D. How does the manipulability ellipsoid change in this case? (Hint: It is not an ellipse anymore.)

6. A manipulability analysis is purely geometrical and depends on joint configuration of a given kinematics. Therefore, it is possible to use this analysis to characterize other (non-traditional) robot "arms" as well. Think about a biomechanical analysis of the human arm: In which configurations do you have maximum manipulability? Which configurations correspond to high exertion (i.e., high "torques"), resulting in small exerted forces on the environment?

5 Grasping

Grasping refers to an activity by which a robot moves or manipulates an object, such as changing its shape or pose. This is typically done by attaching an end-effector (or gripper) that is suitable to perform the task at hand. Grasping has the interesting, and very confusing, property that it is relatively easy in practice but very complicated in theory. Consequently, this chapter describes a variety of strategies that will lead to successful grasps for a wide range of objects, even while it remains difficult to answer questions such as *What makes a good grasp?* or *How does one find good grasps?* in any depth, other than by providing simple heuristics. In this chapter, you will learn the following:

- How to mathematically describe grasping and where simple models reach their limitations
- What the properties of a gripper are that make for a good grasp in practice
- How to understand the trade-offs in a variety of grasping mechanisms

5.1   The Theory of Grasping

Because of the importance of grasping in robotics, the theory behind grasping is widely investigated, with the state of the art comprehensively described in Rimon and Burdick (2019). However, robotics researchers still have difficulties in mathematically capturing the mechanics of grasps that are effective in practice. Therefore, rather than describing these recent developments—which are well beyond the scope of this book—we will briefly describe different approaches to model grasping and their limitations. Our goal is to provide the reader with a better understanding of the reasons why some grasps work better than others and what matters when designing a gripper.

In its most simple form, grasping requires immobilizing an object, at least against the forces of gravity, by providing appropriate forces in the opposite direction, also known as *constraints*. Specifically, contact points on a robotic finger, gripper, or hand are assumed to exert localized forces, thereby constraining the object sufficiently. By this, fingers essentially act as miniature robotic arms, allowing us to apply the methods and tools described in chapters 2 through 4.

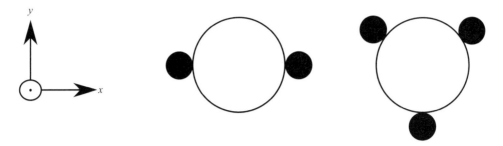

Figure 5.1
Cross-section from above showing an idealized two-finger (left) and three-finger (right) gripper holding a cylinder.

5.1.1 Friction

In any real application, contacts between a gripper and hand are not friction less. This is the reason grasps, such as those shown in figure 5.1, practically work. If there were really no friction between the fingers and the object, the object would be ejected from the hand for every grasp that is not exactly aligned with a principal axis of the cylinder (figure 5.1, left). Furthermore, even the three-finger grasp (figure 5.1, right) would *always* fail as there is no force constraining the object from below. Interestingly, the existence of friction makes grasping much easier in practice yet much harder to describe mathematically.

The reason that the grasps shown in figure 5.1 do work in practice is that the normal forces shown have a tangential component that is due to friction and covered by the *Coulomb's friction law*, which states that the higher the friction coefficient of a material, the more normal force translates into tangential forces that can resist two surfaces from moving against each other. It is governed by the equation:

$$F_t \leq \mu \, F_n. \tag{5.1}$$

Here, F_t is the force of friction exerted by each surface on the other and F_n is the normal force; the force F_t acts in tangential direction to the normal force applied by, for example, a fingertip; μ is a coefficient of friction that can be measured empirically—intuitively, μ is low for glass on glass and high for rubber on wood. We are therefore interested in designing grippers with high-friction coefficients to avoid objects from slipping.

When do objects slip? Let's analyze the problem using figure 5.2. Say we have a fingertip pressing down on a surface in any orientation. There will be a force normal to the surface F_n, which defines the tangential force F_t in any direction. Sweeping the tangential force around the normal force creates a cone with an opening angle of

$$\alpha = 2 tan^{-1} \mu. \tag{5.2}$$

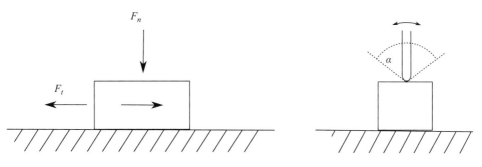

Figure 5.2
Left: Coulomb friction relates normal to tangential reaction forces that are required to overcome friction, here shown for rightward motion. Right: Friction cone for point forces. As long as the force is within the cone, the finger will not slip.

See Rimon and Burdick (2019, 57) for a derivation. If the net force on the object is not within this cone, the object slips. This becomes more intuitive when thinking about how different values of μ affect the shape of this cone. If μ is high, the cone will be relatively wide, letting the object "accept" forces from many different directions without slipping. If μ is low, the cone will be relatively narrow, requiring the force to be normal to the object's surface to prevent slippage.

Importantly, as detailed in chapter 4, a force applied to a rigid body will exert both a three-dimensional (3D) force as well as a 3D moment to the body's center of gravity; this quantity is called a *wrench*—see equations (4.1) and (4.2). If we consider all the possible wrenches that we can apply to a rigid body without having the end-effector slip to form a space (namely, the cone described earlier for a single finger), we can talk about the *grasping wrench space*, which is the corresponding space of all suitable wrenches. Knowing the relation between normal and tangential reaction forces can help in designing grippers that are more likely to successfully grasp an object than others, as well as when planning suitable grasp for objects with known friction.

5.1.2 Multiple Contacts and Deformation

In practice, no force will ever be applied at a single point only; rather, a force will be distributed over an area, either because of the size of the finger pad itself or the contact area deforming under pressure. Even the smallest contact area that is not a point in the mathematical sense will add constraints on torque, which will translate to constraints in additional dimensions and therefore further stabilization of the grasp. This is illustrated in figure 5.3. A single point of contact (figure 5.3, left) allows the object to easily pivot around it; however, by increasing the contact area we are able to constrain the rotational degree of freedom (DoF), thus reducing the available DoFs for the object to only its translational component. It is therefore desirable to grasp an object with a contact area that is as large as possible. Importantly, since surfaces are not perfectly flat, this extension of contact area

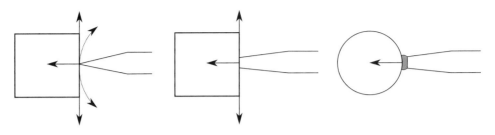

Figure 5.3
From left to right: ideal force exerted via a single point of contact, forces exerted through an area of contact, contact area increasing from pressure and conforming with the surface. Remaining degrees of freedom are indicated by arrows.

is only possible in practice if the contact area is deformable (see figure 5.3, right). Lastly, a large contact area will also increase friction, which as detailed in section 5.1.1 is a desirable property for grasping.

Consequently, using metal jaws or rigid fingers is seldom successful in practice. Instead, rubber pads are used to increase force closure by conforming around the object. As the rubber is flexible, however, the grasp is not completely stabilizing the object and it may still be able to move within the grasp; this might not be desirable for precise manipulation tasks such as picking up a nut and trying to mount it on a screw. Mathematically, this introduces additional complications into the grasp model in the form of elasticity in the object–robot dynamics; in simple terms, soft/flexible pads are the equivalent of a spring, increasing uncertainty in the model's dynamics.

5.1.3   Suction

A highly effective method for grasping is using suction. Here, a suction cup is pressed against an object, using a vacuum applied by a pump to suck the object against the cup. Instead of exerting forces against the object, which always requires at least one antipodal force (or multiple forces that are distributed such that the object remains in equilibrium) to create a constraint, suction only requires one point of contact. The rim of the suction cup provides both friction and multiple contact points to prevent the object from slipping and further constraining the object beyond the normal force applied by the vacuum. Requiring only a single area of contact is a tremendous advantage from a planning perspective because only one area on an object needs to be identified, whereas other grasping approaches need to always identify two areas and coordinate motion to reach them. It is worth noting that suction using multiple suction cups on custom-made rigs to grasp large parts such as car doors is very popular in the car industry, but it generally relies on preprogrammed trajectories and little to no autonomy—which is not a focus of this book.

The soft nature of the suction cup allows the rim to conform to the object, but it makes suction impractical for objects that do not have any flat surfaces or that expose holes to the gripper—for example, objects stored in a net. The elasticity of the rim also makes it difficult to further manipulate the object as all forces applied by the robot will need to be transferred through a spring like elastic material. Finally, suction requires a vacuum pump that is able to generate sufficient force to lift an object, limiting the maximum weight of objects suitable for suction by a single suction cup in practice.

5.2 Simple Grasping Mechanisms

Understanding why grasping actually works—namely, by friction (section 5.1.1) and increasing contact area from deformation (section 5.1.2)—allows us to select grasping mechanism that are characterized by the following properties: (1) They are able to successfully grasp a wide range of objects, (2) they are simple to construct, and (3) they are easy to control. Here, properties of interest are the range of possible object sizes, the maximum weight of an object, and how fragile objects can possibly be. Object dimensions are directly dependent on the gripper kinematics, such as minimum and maximum aperture, whereas the maximum weight is given by the torque the mechanism can exert as well as the number of contacts and their friction parameters. Whether a gripper can handle fragile objects is a function of how well this torque can be measured and controlled.

5.2.1 1-DoF Scissorlike Gripper

One of the simplest grippers is a simple one degree-of-freedom (1-DoF) claw, which is a popular design in the prosthetic community and has been refined for centuries. Actuated by a string mounted to a person's shoulder, or more recently by electric motors controlled by muscle activity in the lower arm, this simple mechanism allows the wearer to perform a wide range of everyday activities. Indeed, an off-the-shelf prosthetic hand has been shown to perform a large variety of grasping and manipulation tasks when compared with other robotic hands in a teleoperation scenario, limited only by its ability to conform to specific kinematic constraints such as operating scissors (Patel, Segil, and Correll 2016).

A simple design is shown in figure 5.4 and consists of an active finger that presses an object against a passive finger, with both fingers often shaped as a hook. As should be clear by now, such a design can only work by relying on friction, which makes it not very common in traditional robotics.

The key advantage of this mechanism is that it allows for very simple control strategies to operate it: Use the passive finger to make contact with the object, then use the active finger to close the grasp. The event "make contact" can either be detected by measuring the force acting at the wrist and looking for abrupt changes in such force or using a tactile sensor on the finger with which contact is made. This approach can therefore lead to robust grasps with a minimum of sensing required. A disadvantage of this mechanism is that its function relies exclusively on friction, possibly ejecting away objects from its grasp if friction is

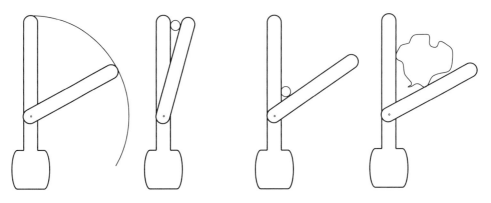

Figure 5.4
Simple 1-DoF grasping mechanism that relies on friction to grasp objects with a wide variety of sizes (center, right). The mechanism has only one moving part that presses the object against a passive finger.

not sufficient or the object is in an otherwise suboptimal conformation. Unlike most other mechanisms, it is also impossible to use the finger position to infer the width of an object, (figure 5.4, center).

The mechanism shown in figure 5.4 can be actuated in many different ways—for example by attaching a servomotor directly to the active finger, using a shape-memory alloy wire with a suitable lever arm, or using a pneumatic piston or balloon.

5.2.2   Parallel Jaw

The most common industrial gripper mechanism is the two-finger parallel jaw gripper. It operates by squeezing an object between its two parallel jaws, which are usually driven by a single actuator and therefore move in concert. Parallel jaw grippers usually yield more contact area than a scissor-like 1-DoF gripper, but they suffer from a smaller range of motion.

The left side of figure 5.5 shows a minimalist implementation of a parallel jaw gripper that can be actuated by a single servomotor, driving two rack gears to which the gripper jaws are mounted. While using gears on racks is unusual in an industrial design—the gripper jaws typically travel on threads actuated by worm gears or are attached to a pneumatic piston— figure 5.5 illustrates the relationship between the range of motion of the gripper jaws, the length of the mechanism it is sliding on (here a rack gear), and the resulting body size. In order for this design to fully close, the two rack gears must be mounted at an offset in order to slide against each other. Constraints like this often make the gripper body twice as wide as the maximum aperture, making it difficult for the robot to enter tight areas. The mechanical design also affects the speed at which a gripper can operate. Pneumatic grippers, where air pressure coming in on either end of the piston can drive the gripper into an "open" or "close" position very quickly (two or three times per second), cannot

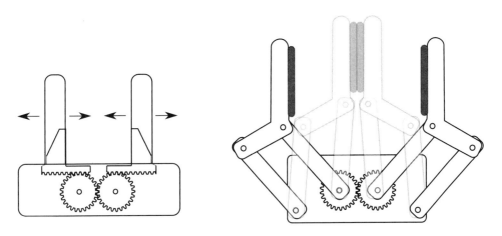

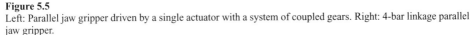

Figure 5.5
Left: Parallel jaw gripper driven by a single actuator with a system of coupled gears. Right: 4-bar linkage parallel jaw gripper.

be controlled accurately. Electric mechanisms instead trade-off accuracy and torque with speed (i.e., more accuracy but at lower speed).

The control strategy for parallel jaw grippers requires an accurate pose estimation of the object of interest and precise positioning of the gripper so that the object is right in the center of the two jaws. Note that force closure with a static object, such as a screw mounted to a structure, requires both jaws to make contact with the object at the same time, thereby imposing high accuracy requirements on both object detection and robot motion. Here, compliance can help, allowing the gripper to adjust its pose to the object. This can be accomplished by either measuring forces in the wrist and moving the gripper to minimize lateral forces or a compliant mounting mechanism or structure, such as a robot equipped with series-elastic or pneumatic actuators. An alternative approach is to actuate both gripper jaws independently.

5.2.3 4-Bar Linkage Parallel Gripper

A parallel jaw mechanism with a larger range of motion can be accomplished using two 4-bar linkages (figure 5.5, right). In a 4-bar linkage, rotation of the motor is translated into straight translation of the fingers. This is accomplished by two pairs of parallel bars of equal lengths. Although one might think that the mechanism is only made out of three bars when inspecting figure 5.5, the gripper body itself takes the role of the fourth bar. Interestingly, both pairs remain parallel as one of the bars is rotating, resulting in the two gripper jaws remaining parallel to each other. This is best understood by inspecting figure 5.5 and comparing the two positions the left jaw can be in.

The drawback of this design is that closing the gripper also results in a forward motion. This requires approaching an object from different distances, depending on its width. Other than this, the control strategy is the same as for the parallel jaw gripper, requiring an accurate estimate of the object's pose. Also here, adding compliance or independent actuation of each jaw can help resolve accuracy problems.

5.2.4   Multifingered Hands

Grippers with more than two fingers/jaws are rarely used in industrial practice. One common use case is grasping cylindrical objects from above; in this case, three-fingered hands (see figure 5.1, right) are best suited. However, in most other cases three fingers are not an advantage and might even be a hindrance! For example, it is difficult to perform simple pinching grasps with three fingers. This has led to designs in which two of the fingers are reconfigured from performing an inward motion to behave identical to a parallel jaw gripper, while the third finger is stored in a safe position. In addition to mechanical complexity, such an approach requires extra planning steps.

   How many grasps are possible and how many possible grasps are needed to grasp every possible object remains a difficult theoretical problem (which is further complicated by the fact that successful grasping often happens at the boundary of what is mathematically tractable). Generally, we can say that additional fingers—such as in the human hand—provide additional redundancy, which allows grasping and manipulating (see chapter 14) the same object in many different ways, including manipulating the object within the hand—that is, without intermittent placement or handing it over to another gripper.

Take-Home Lessons

• Making a good gripper requires taking advantage of compliance and friction in a way that still eludes mathematical analysis, making gripper design an experimental process.

• Successfully grasping an item does not necessarily mean that the robot will also be able to successfully manipulate that item. Designing a grasping mechanism therefore requires understanding the entire task, from grasping to placing or otherwise manipulating the item.

• Simple mechanisms such as suction or two-finger grippers are sufficient for most grasping and manipulation tasks, but they are not suitable for in-hand manipulation, which—for the large part—remains an open research challenge.

Exercises

1. Think about at least three mechanisms to realize a parallel jaw gripper. How does the minimum and maximum aperture of the gripper relate to the gripper width for each of these designs?

2. Think about at least three mechanisms to actuate a 4-bar linkage. Which of these will keep the payload inside the gripper during power failure?

3. Derive an equation for the distance of the fingertip from the gripper base in a 4-bar linkage gripper as a function of the gripper opening width. Use appropriate parameters for all unknown parameters.

4. A powerful mechanism to grasp is to evacuate the air from a bag of coffee beans after conforming around an object. Describe what happens here using language from this chapter.

5. Design a grasping system to pick up the bare metal of a car door for assembly in an automated manufacturing line. Which design ensures maximal accuracy in placing the part?

II SENSING AND ACTUATION

6 Actuators

The first part of this book has been concerned with different mechanisms, helping us understand what robots look like and how they move. We now introduce the devices that turn energy into rotation and translation. We generally call such devices *actuators*. The goals of this chapter are to provide readers with the following foundational information:

- A general overview of the different types of actuators and their advantages and drawbacks
- An appreciation of the systems challenges that come with any chosen actuator technology
- A basis and reference for further study

6.1 Electric Motors

Because of the dominance of rolling robots, the electric motor (Hughes and Drury 2019) is among the most popular actuators. Electric motors come in different variations, starting with stepper and DC motors to servo and so-called brushless DC motors. Except for the stepper motor, which uses large electromagnets to rotate an internal spindle by a few degrees every time, the physics of the electrical motor requires it to revolve at very high speeds (multiple thousand rotations per minute). Therefore, electric motors are almost always used in conjunction with so-called reduction gears tasked with reducing their rotational speed and increasing their torque (i.e., the rotational force that the motor exerts to rotate about its axis). Torques are indeed one of the most basic control commands that may be issued to control a motor (i.e., the lowest-level). A notion of joint torque was introduced in section 4.1 and is generally employed when controlling a robot at the static or dynamic levels of abstraction. In order to be able to measure the number of revolutions and the axis position, motors are also often combined with rotary encoders (section 7.2). Motors that combine an electric motor with a gearbox, encoder, and controller to move toward desired position are known as servomotors, and they are popular among hobbyists.

(a) (b)

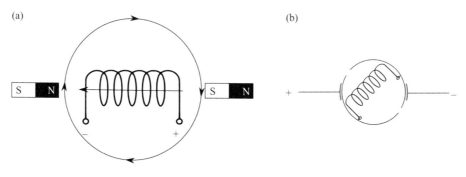

Figure 6.1
(a) A simple motor consisting of a coil and two permanent magnets. In its current configuration, the electromagnetic force will result in the coil turning 180 degrees. (b) A simple commutator, which will switch the direction of the current and therefore of the magnetic field as the coil rotates.

6.1.1   AC and DC Motors

Electric motors turn electric energy into kinetic energy by means of electromagnetism. More specifically, an electric current running through a wire creates a circular magnetic field around the wire according to Ampere's law. This effect can be amplified by winding the wire into a coil. Because of the coil shape, the magnetic fields of all wires superimpose and create a strong magnetic field in the center of the coil. This field can be used to magnetize a ferromagnetic material such as iron, which in turn amplifies the magnetic field. The resulting *magnetomotive force* (MMF) is directly proportional to the number of windings of the coil and the current running through it.

AC motors
In an AC motor, an electromagnetic coil is usually paired with a permanent magnet. As magnets of opposite polarity repel each other, this effect can be used to create motion. In its simplest form, an electric motor consists of a simple coil that can spin between two permanent magnets, one oriented with its south pole to the center, the other with its north pole (figure 6.1a). Attaching a shaft to the central coil would then make it possible to turn a wheel, for example. The turning part with the shaft is known as the *rotor*, whereas the static part is known as the *stator*. When running an electric current in the coil, the iron core will get magnetized: Its north pole will then be attracted by the south pole of one magnet in the stator and repelled by the north pole of the other permanent magnet in the stator, while the opposite will happen to its south pole.

In order for this simple motor to not get stuck in its new configuration, we will need to swap the direction of the rotor's magnetic field. This can be achieved by swapping the direction of the current running through the coil. This happens by itself when using so-called *alternating current* (AC). AC is commonly used in the power grid, where the direction of current changes with a frequency of 50 or 60 hertz (Hz). In this case, the speed of the motor depends on the frequency of the AC, whereas its maximum torque is given by its current.

AC motors exist in different forms, often using multiple coils in parallel pairs to create smoother motions. Some motor designs also place the permanent magnets onto the rotor and coils on the outside. However, whatever the design is, the basic principles remain the same.

DC motors

As the speed of an AC motor is constant, it is mostly used in heavy industrial applications. An alternative design is to generate the desired switch in directionality by a so-called *commutator* (figure 6.1b). This allows running the motor with what is known as *direct current* (DC), in which the direction of current does not change. DC is what is commonly available from batteries or from a "wall wart" that converts AC into DC by means of a transformer and a rectifier.

The commutator now provides positive and negative voltage at a series of interleaving pads. These can be placed along the circumference of the stator and provide power to the rotor coil through metal brushes attached to the shaft. By this, the central coil will receive power at the right polarity no matter where it is. As with the AC motor, there exist multiple designs using pairs of parallel coils, various brushes, and commutators mounted either in the stator or the rotor.

Such a motor can now turn at arbitrary speeds and will become faster and faster, only limited by friction and torque applied to its shaft. Its speed is therefore proportional to the voltage that is applied, whereas its torque is limited by the maximum current that is provided.

Electric DC motors are widely used in robotics but suffer from low efficiency because of the friction of the brushes and their wear and tear.

6.1.2   Stepper Motor

Even when using more than one coil and multiple pairs of permanent magnets, it is difficult to precisely control the angular position of a DC motor shaft. Although the rotation can be geared down by factors of hundreds or even thousands, the motor itself usually spins in the order of thousands of times per minute—also known as "rotations per minute" (RPM). A solution to this problem is the *stepper motor* that—in its simplest form—uses a ferromagnetic rotary wheel with a fixed number of teeth as its stator. Coils in its stator can attract these teeth, creating a small rotation of a few degrees when the teeth in the rotor and stator-coil align. To precisely control this effect, the ferromagnetic material in the coil also has a teeth pattern. Selectively turning pairs of coils on and off will allow the motor shaft to turn at a fixed number of degrees (in the order of one degree or less). For example, a stepper motor that turns 3.6 degrees per step will require 100 steps for a complete revolution.

The required voltage pattern is usually generated by a microcontroller. A stepper motor with four phases—that is, four sets of coils inside—requires four electrical signals that are carefully interleaved. The first wire is on for a set amount of time while the other three are off, then the second, the third, and the fourth. Here, the period (i.e., the length) of this signal determines the stepper motor's speed, whereas the maximum current determines its holding torque. There are a variety of low-cost integrated circuits (ICs) that generate this pattern, reducing the microcontroller's task to simply sending a single bit for every step and another for the desired direction.

The advantage of this approach is that stepper motors usually do not require gears or encoders (as one can simply count the steps being sent), making them attractive as drivers for small differential-wheel robots or grippers. Stepper motors are usually much more expensive and bulky than their DC counterparts.

6.1.3   Brushless DC Motor

As the alternating current patterns are generated by electronics, the stepper motor does not require brushes to commutate and is therefore much more efficient. The advent of micro-electronics in the 1970s made it possible to generate driving patterns at the speeds required by conventional DC motors (thousands of RPMs), which has led to the *brushless DC motor*. The brushless DC motor indeed resembles a stepper motor, but it can operate with much smaller coils because its torque results from the kinetic energy of rotating at high speeds. In order to improve control, brushless DC motors either use encoders, a Hall effect sensor, or small changes in current (resulting from the dynamo effect in the currently unused coils) to measure the current position of the rotor within the stator. Sensing and control of a brushless DC motor is involved and usually provided by purposely designed solid-state electronic devices.

Unlike a brushed DC motor, whose brushes induce friction, the maximum speed of a brushless DC motor is mostly limited by heat, which is a by-product of running electric current through its coils. Because of the absence of friction, brushless DC motors are far more efficient than their brushed counterparts and can provide equivalent speed and torque at a smaller form factor and lower weight.

The performance of electric motors has been further boosted by the discovery of rare earth magnets such as neodymium in the 1980s, allowing motors to exert even more torque at smaller weight and using lesser current. Together, these advances have led to a renaissance of electric cars and, together with solid-state inertial measurement units (IMUs) (section 7.3.2), enabled small-scale drones.

6.1.4   Servo Motor

To be useful in a robotic system, electric motors usually require gears, an encoder, and control electronics. Modules that package these components into a convenient form factor are known as *servomotors*.

Servomotors have been classically used in remote-controlled cars to provide a simple actuator to steer a car or move the flaps of an airplane. A simple digital signal was used to set the servo angle, usually in the range of 360 degrees or less, which was then held by the integrated electronics. More recently, newer digital servomotors have emerged that are able to not only control the angle but set the speed at which the servo moves and the maximum current (and thereby torque). They can also read information such as actual angle, temperature, and other operational parameters.

Because of their built-in gear reduction, servomotors are usually not suitable in the drive-train of mobile platforms, but they have become increasingly prominent to drive the joints of simple manipulating arms, articulated hands, and grippers. A special kind of servomotor is the *linear actuator*. Here, a (brushless) DC motor is driving a spindle that turns rotation into translation. Linear servomotors are available with a wide range of protocols and with or without built-in encoders that provide position feedback.

6.1.5 Motor Controllers

Designing the power electronics that turn digital information into precisely controlled voltage and currents is particularly challenging. Transistors are used to turn low-power control signals from a microcontroller into high-powered ones, while regulating the voltage is achieved by switching DC power on and off at very high frequency—tens of kilohertz (kHz)—and smoothing the signal using a combination of capacitors and coils. Diodes are used to ground the reverse voltage that arises from demagnetizing coils. As peak loads of tens of amperes will already arise in smaller systems such as remote-controlled cars, designing a motor controller is very involved and usually limited to a specific operational range.

The biggest challenge in selecting appropriate circuits for motor drivers is selecting a circuit that can not only accommodate the voltage (U) and current (I) requirements but can actually handle the overall energy ($P = UI$). Here, the first hurdle is to provide the desired supply voltage. As it is difficult to convert supply voltages without loss (in particular when the required current is large), voltage requirements of the main driving motors often determine the operating voltage of the overall power system.

A second hurdle is that there is nothing like loss less energy switching. In particular, all power transistors have an internal resistance (R). Given $P = I^2R$, already small resistances generate substantial heat. Dissipating the resulting heat can quickly become a major problem; typically, this is not part of an off-the-shelf motor control solution, but it represents a mechanical design problem in and of itself. A standard approach is to use the (metal) robot chassis to dissipate heat, but sometimes active cooling using a fan is necessary. For more details on designing power electronics for a variety of electric motors, the reader may refer to Hughes and Drury 2019.

6.2 Hydraulic and Pneumatic Actuators

Another popular class of actuators, in particular for legged robots, are linear actuators that might exist in electric, pneumatic, or hydraulic form.

6.2.1   Hydraulic Actuators

Hydraulic actuators, mostly in the form of pistons, are well known from construction machines and other heavy equipment. Hydraulics usually exceed the forces electric motors can generate and are in a different ballpark as far as size is concerned. That is, the smallest available hydraulic actuators are orders of magnitude larger (on the order of tens of centimeters) than the smallest DC motors (on the order of millimeters). However, they are relevant for larger bipedal and quadrupedal platforms, where they are often used in conjunction with electric motors.

Hydraulic actuators require a tank with pressurized liquid and a compressor pump (which is again driven by a DC motor). The liquid is pressurized using a gas, and released into the actuators via solenoid valves. A second solenoid valve is used to let the liquid escape the actuator. The compressor is then pumping the liquid back into the tank. Here, the gas in the tank acts as a buffer, allowing the system to release a burst of energy, which then needs to be slowly restored by the compressor. As the performance of a hydraulic system is strongly related to its mechanical properties such as size and pressure of the tank, diameter of the tubes connecting the components, and dimensions of the valve, hydraulic systems have a narrower operational range than electric motors, which allow a higher variation of forces and speeds. However, they are costly to maintain, difficult to control, and usually characterized by a low bandwidth: that is, they will never be as reactive as electric motors, and they might be infeasible in human-populated environments where speed of reaction is paramount for safety considerations (section 6.3).

6.2.2   Pneumatic Actuators and Soft Robotics

The principles of fluid-based (hydraulic) actuators also extend to operation by air. Pneumatic systems also require a compressor, a tank, and a set of valves to direct the flow of air. Because air is in orders of magnitudes more compressible than a liquid, pneumatic systems are not well suited to translate large forces; however, they are lightweight and available in much smaller form factors than hydraulic systems. For example, solenoid valves can be as small as a few millimeters, so it is possible to construct intricate mechanisms such as realistically sized robotic fish (Katzschmann et al. 2018) or robotic hands (Deimel and Brock 2016).

In addition to pistons and other actuators available for hydraulic systems, pneumatic actuators can be designed in arbitrary form factors, allowing the designer to turn air pressure into almost any desired bending or torsional movement. These actuators usually consist of a flexible rubber material with an internal cavity that can be filled with air. Materials such as fabrics that are stiff in one dimension (when pulling) but flexible in another (when bending) are used to direct the force of the incoming air into a desired direction, resulting in the actuator bending or twisting in a desired way (Polygerinos et al. 2017). Please note that soft actuators are *not* balloons. While balloons change their volume as they are inflated, a change of volume is considered a failure mode in a soft robotic actuator and, ideally, all energy should directly convert into motion.

As soft actuators are flexible, they break with the tradition of kinematics for rigid robots introduced in chapter 3. From a kinematic perspective, an ideal, fully soft robot can be modeled as a platform with an infinite number of mechanical degrees of freedom! However, although their complex mechanics makes modeling and control more difficult, soft robotics opens up an entirely new spectrum of robot capabilities; for example, it is possible to design nontraditional kinematics that more resemble the motion of animals than those of machines. This also makes it possible to employ control strategies that rely on the mechanism known as *compliance*, a concept briefly introduced in section 5.2 on robotic grasping and enabled by force control (chapter 4).

6.3   Safety Considerations

Which actuator system is the best choice is also driven by safety considerations. We distinguish *active* and *passive* safety. Active safety is the ability to control the actuator sufficiently to prevent it from doing harm—for example, by squeezing a person's finger or limb or damaging infrastructure in the environment. The class of *collaborative robots* achieves that by limiting the torque a robot can achieve through control. Controlling torque can be accomplished by measuring and regulating the current that is used at any given actuator and employing a suitable low-level dynamical model of the motor that relates operating current with motor torque. By comparing the joint torques needed to perform a task (see chapter 4) with the currents actually flowing through the motors, a robot can detect whether it is about to exert a (potentially harmful) force or torque on the external world. This usually requires estimating the approximate weight of the robot's payload. A better (but more expensive) way to control torque is to measure it at each joint by means of load cells (see chapter 7). Using actual sensors to measure forces and torques actually helps to extend this method to other actuation systems, not limited to electrical motors.

Passive safety is the ability to maintain a robot's safety even in the absence of control. This can be achieved by cushioning a robot, using actuators that "give in" at a lower force than can do harm, such as pneumatic actuators, or by coupling the motor with an elastic element. In such *series-elastic actuators*, power is not directly transmitted by a motor shaft but indirectly through a spring. In addition to doubling as a force/torque sensor (see also chapter 7), this limits the maximum force that such an actuator can exert and reduces the precision of any high-level controller.

An important failure mode that can only be addressed by a passive safety mechanism is *power failure*. In the absence of power, a mobile robot might keep driving (based on inertia) and hit an object, a humanoid robot might collapse onto itself, and a gripper might let a heavy (or sharp) payload fall. These problems can be addressed with always-on breaking mechanisms that engage in the absence of power or by using gear ratios that are so high that the overall mechanism is not back-driveable.

Take-Home Lessons

• There exist an almost limitless repertoire of techniques to turn energy into motion, many of which have been explored to create robots, with the electric motor remaining the dominant actuator in small-scale robotic systems.

• What makes a good actuator is determined by its efficiency with respect to the available energy source, as well as how far its position, velocity, and torque can be measured to enable accurate and precise control.

• A robot's safety is determined not only by the choice of actuator but also by the control system around it.

Exercises

1. You are designing a robotic arm. Your goal is to maximize strength while minimizing weight. Which kind of electric motor do you chose and why?

2. You are designing a gantry system for a 3D printer that can move the printhead left, right, up, and down.

a) What kind of electric motor is your preferred choice and why?

b) The motor you selected requires 5 volts and up to one ampere. Select an appropriate driver board from an online vendor.

c) What additional components would you need to realize the gantry with a brushless DC motor?

3. A motor you selected for the shoulder joint of a robotic arm is too weak and stalls under load. Assuming power is provided uninterrupted, what will eventually lead to permanent damage?

4. The key component in a motor controller has a so-called ON-resistance of 0.2Ω. Your motor requires 10A at average load.

a) What is the power dissipated via heat?

b) Search the internet for thermal heat sinks. What are the key quantities to look out for here?

5. Compare the parallel jaw gripper with a 2-bar linkage gripper. Discuss their safety properties in case of power failure.

6. The end-effector of a "soft" robotic arm is hitting an object with with a velocity of 3m/s. Discuss its safety when compared with a conventional robotic arm and state the assumptions you are making.

7 Sensors

Robots are systems that sense, actuate, and compute. So far, we have studied the basic physical principles of motion—namely, locomotion and manipulation. We now need to understand the basic principles of robotic sensors that provide the necessary data for a robot to make decisions and control itself. The goals of this chapter are to:

- Provide an overview of sensors commonly used on robotic systems.
- Outline the physical principles behind the functioning of sensors.
- Clarify the mechanisms responsible for uncertainty in sensor-based reasoning.

Historically, the development of robotic sensors is driven by industries other than robotics, including transportation (automotive, naval, airplanes), safety devices for industry, servos for remote-controlled toys, and more recently cellphones, virtual reality, and gaming consoles. These industries are mostly responsible for making "exotic" sensors available at low cost by identifying mass-market applications. For example, accelerometers and gyroscopes are now widely used in smartphones and cost less than a dollar; the Xbox gaming console made three-dimensional (3D) depth sensing (through the Microsoft Kinect) accessible for a greatly lower cost than before; and sensors in modern passenger vehicles provide an array of capabilities without appreciably increasing the cost of the vehicle itself.

> Think about the sensors that you are interacting with daily. What sensors do you have in your phone, in your kitchen, or in your car?

As we will see later on, sensors are hard to classify by their application domain and target use case. In fact, most problems benefit from every possible source of information they can obtain. For example, localization can be achieved by measuring how many degrees a wheel has turned with a sensor known as an "encoder" (discussed in section 7.2) that can be implemented relying on a wide variety of physical principles. However, estimation becomes more precise with the addition of accelerometers (section 7.3) or even vision sensors (chapter 8). All of these approaches differ drastically in their precision—a term that will be more formally introduced below—and the kind of data they provide, but none of them is able to completely solve the localization problem on its own.

> Think about the kind of data that you can obtain from an encoder, an accelerometer, or a vision sensor on a non-holonomic robot. What are the fundamental differences? What physical principles do they leverage?

Although an encoder is able to measure position, it is used in this function only on robotic arms. If the robot is non-holonomic, closed paths in its configuration space (i.e., robot motions that return the encoder values to their initial position) do not necessarily drive the robot back to its starting point (as exemplified in figure 3.4). In those robots, encoders are therefore mainly used to measure speed. An accelerometer instead, by definition, measures the derivative of speed. Vision, finally, allows for the calculation of the absolute position (or the integral of speed) if the environment is equipped with unique features. An additional fundamental difference between those three sensors is the amount and kind of data they provide. An accelerometer samples real-valued quantities that are digitized with some precision. An odometer instead delivers discrete values that correspond to encoder increments. Finally, a vision sensor delivers an array of digitized real-valued quantities (namely colors). Although the information content of this sensor exceeds that of the other sensors by far, cherry-picking the information that is the most useful to complete a task remains a hard and generally unsolved problem.

7.1   Terminology

When dealing with sensors, it is important to provide precise definitions of terms such as "speed" and "resolution" as well as additional taxonomy that is specific to robotics. Roboticists differentiate between *active* and *passive* sensors. Active sensors emit energy of some sort and measure the reaction of the environment. Passive sensors instead measure energy from the environment. For example, most distance sensors (not including stereo vision) are active sensors because they sense the reflection of a signal they emit; conversely, an accelerometer, a compass, or a push-button are passive sensors. Frequently the addition of an active element to a passive sensor can increase the signal-to-noise ratio of the passive sensor, so these distinctions may be blurred in some cases.

Another important term to characterize sensors is its *range*, which is the *difference* between the upper and the lower limit of the quantity a sensor can measure. This differs from its *dynamic range*, which is the *ratio* between the highest and lowest value a sensor

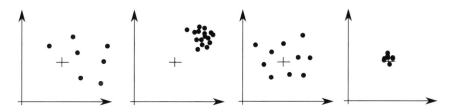

Figure 7.1
The cross corresponds to the true value of the signal. From left to right: neither precise nor accurate, precise but not accurate, accurate but not precise, accurate and precise.

can measure. It is usually expressed on a logarithmic scale (to the basis 10), also known as "decibel." The minimal distance between two values a sensor can measure is known as its *resolution*. The resolution of a sensor is primarily limited by the physical principle it leverages (e.g., a light detector can only count multiples of a quant), however it is usually limited by the analog-digital conversion process. The resolution of a sensor should not be confused with its accuracy or its precision (which are two different concepts). For example, an infrared distance sensor might produce 4,096 different values to encode distances from 0 to 10 centimeters which suggests a resolution of around 24 micrometers; even so, its precision is much lower than its resolution (usually on the order of millimeters) due to noise in the acquisition process.

A sensor's *accuracy* is the difference between its (average) output m and the true value v to be measured:

$$accuracy = 1 - \frac{|m - v|}{v}. \tag{7.1}$$

This measure provides a quantity that approaches 1 for very accurate values and 0 if the measurements group is far away from the actual value. In practice, however, this measure is rarely used, and accuracy is provided with absolute values or a percentage at which a value might exceed the true measurement.

A sensor's *precision* instead is given by the ratio of range and statistical variance of the signal. As detailed in figure 7.1, precision is therefore a measure of *repeatability* of a signal, whereas accuracy describes a *systematic error* that is introduced by the sensor's physics. For example, a GPS sensor is usually precise within a few meters but only accurate to tens of meters. This becomes most obvious when satellite configurations change, resulting in the precise region jumping by a couple of meters. In practice, this can be avoided by fusing this data with other sensors (e.g., from an inertial measurement unit).

The speed at which a sensor can provide measurements is known as its *bandwidth*. For example, if a sensor has a bandwidth of 10 hertz (Hz), it will provide a signal ten times a second. This is important to know, as querying the sensor more often is a waste of computational time and potentially misleading.

7.1.1 Proprioception versus Exteroception

Another important taxonomy is the difference between proprioception and exteroception. *Proprioception* refers to the perception of the internal state of a robot. It includes an estimation of the robot's joint angles, its speeds, as well as internal torques and forces. Conversely, *exteroception* refers to sensing anything outside the physical embodiment of the robot. Exteroception is important because it is crucial for the robot to correctly perceive the state of the world, estimate the uncertainties related to it, and properly act based on these uncertainties. Importantly, while the majority of sensor development focuses on *distal* sensors capable of measuring quantities in the far space (e.g., cameras, as discussed in chapter 8, or sound-based sensors, covered in section 7.5.3), in recent years more attention has been given to *proximal* sensors that are concerned with measuring the environment that is immediately surrounding the body or even directly on the robot body. Applications of this technology are varied, from tasks that require measuring and controlling the interaction of the end-effector with the environment (e.g., sanding a table with a fixed vertical force) to manipulating in clutter—where contact with obstacles is inevitable.

> In robotics, it often helps to make comparisons with human performance. How many daily tasks *do not* require physical interaction with the environment? If they do, would you be able to successfully complete them without contact, and how would your performance decrease if you were to "avoid collisions at all costs"?

7.2 Sensors That Measure the Robot's Joint Configuration

The most important proprioceptive sensor is the *encoder*. Encoders can be used for sensing joint position and speed as well as force—if used in conjunction with a spring. Encoders can be divided in incremental (relative, used primarily in mobile robotics) and absolute encoders (used mainly in robot manipulators). In general, they rely on either a magnetic or optical beacon turning together with the motor and being sensed by an appropriate sensor that counts every pass-through. The most common encoder in robotics is the *quadrature encoder*, which is an optical encoder. It relies on a pattern rotating with the motor and an optical sensor that can register black/white transitions, as shown in figure 7.2.

While a single sensor would be sufficient to detect rotational position and speed, it does not allow one to determine the direction of motion. Quadrature encoders therefore have two sensors, A and B, that register an interleaving pattern with distance of a quarter phase. If A leads B, the disk is rotating in a clockwise direction. If B leads A, then the disk is rotating in a counterclockwise direction. It is also possible to create absolute encoders—an example of which is shown on the right in figure 7.2. This pattern is a three-bit pattern that encodes

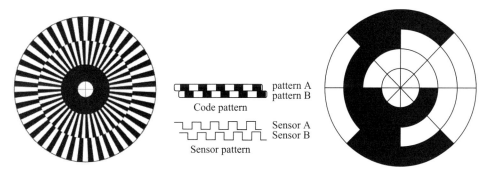

Figure 7.2
From left to right: encoder pattern used in a quadrature encoder, resulting sensor signal (forward motion), absolute encoder pattern (gray coding).

eight different segments on a disc. Notice that the pattern is arranged in such a way that there is only one bit changing from one segment to the other. This is known as "Gray code."

7.3   Sensors That Measure Ego-Motion

Measuring the robot's joint configuration is limited to static observations. It does not allow the robot to detect whether it is currently moving or even accelerating (such as falling), which is particularly important for robots such as walking humanoids or quadrotors that are only dynamically stable. Motion can be estimated by relying on the principle of *inertia*. A moving mass does not lose its kinetic energy—if there is no friction. Likewise, a resting mass will resist acceleration. Both effects are due to inertia and can be exploited to measure acceleration and speed.

7.3.1   Accelerometers

An accelerometer can be thought of as a mass on a dampened spring. Considering a vertical spring with a mass attached to it, we can measure the acting force $F = kx$ (Hooke's law) by measuring the displacement x that the mass has exerted on the spring. Using the relationship $F = ma$, we can now calculate the acceleration a on the mass m. On earth, this acceleration is roughly $9.81 \frac{m}{s^2}$. In practice, these spring/mass systems are realized using microelectromechanical devices (MEMS), such as a cantilevered beam whose displacement can be measured using a capacitive sensor. Accelerometers measure up to three axes of translational accelerations. Inferring an absolute position from it requires a double integration, which introduces significant noise in the estimation and makes position estimates using accelerometers infeasible in practice. However, as gravity provides a constant acceleration vector, accelerometers are very good at estimating the pose of an object with respect to gravity (i.e., roll and pitch).

7.3.2   Gyroscopes

A gyroscope is an electromechanical device that can measure rotational speed and, in some configurations, orientation. It is complementary to the accelerometer that measures translational acceleration. Classically, a gyroscope consists of a rotating disc that can freely rotate in a system of pivots and gimbals so that when it is moving, the inertial momentum maintains the original orientation of the disc. In this way it is possible to measure the orientation of the system relative to where the system was originally. While disc-based gyroscopes are still used, for example, to stabilize the cannon of a tank during motion, the mechanism remains difficult to minimize.

A variation of the gyroscope is the rate gyro, which measures rotational/angular velocities. A rate gyro can intuitively be illustrated by considering its *optical* implementation. In an optical gyro, a laser beam is split in two and sent around a circular path in two opposite directions. If the device is rotated against the direction of one of these laser beams, one laser will have to travel a slightly longer distance than the other, leading to a measurable phase shift at the receiver. This phase shift is proportional to the *rotational speed* of the setup. As lights with the same frequency and phase will add to each other and lights with the same frequency but opposite phases will cancel each other, light at the detector will be darker for high rotational velocities. Importantly, small-scale optical rate gyros are not practical, but MEMS rate gyros are widely available and use a different technology, since they rely on a mass suspended by springs. The mass is actively vibrating, making it subject to Coriolis forces when the sensor is rotated. Coriolis forces can be best understood by moving orthogonally to the direction of rotation on a vinyl disc player. In order to move in a straight line, you will not only need to move forward but also sideways. The necessary acceleration to change the speed of this sideways motion is counteracting the Coriolis force, which is both proportional to the lateral speed (the vibration of the mass in a MEMS sensor) and the rotational velocity that the device wishes to measure. Note that the MEMS gyro would only be able to measure acceleration if it were not vibrating.

Rate gyroscopes can measure the rotational speed around three axes, which can be integrated to obtain absolute orientation. As an accelerometer measures along three axes of translation, the combination of both sensors can provide information on motion in all six degrees of freedom. Together with a magnetometer (compass), which provides absolute orientation, this combination is also known as an *inertial measurement unit* (IMU). Note that this combination of sensors is particularly powerful, as an accelerometer and gyroscope can provide complementary information on roll and pitch while a magnetometer and gyroscope can provide complementary information on yaw. This innovation has powered attitude and heading reference systems (AHRS) through sensor fusion, a technique that is explored in section 7.6.

7.4   Measuring Force

The measurement of physical interaction forces is of paramount importance for robotics. It enables a variety of capabilities that humans take from granted, from gently picking a strawberry to safely engaging in touch-based interactions with humans.

When combining a motor and an encoder with a spring, a mechanism known as a *series-elastic actuator* (Pratt and Williamson 1995), rotary and linear encoders can be used as simple force or torque sensors using Hooke's law ($F = kx$, where k is a spring constant and x is the displacement in the spring from extension or compression). This can be used when operating a robot under a static (section 4.1) or dynamic level of abstraction. Another method to estimate the actual force or torque acting on a joint is to measure the current consumed at each joint. Knowing a mechanism's pose allows one to calculate the resulting forces and torques across the mechanism as well as the currents required for empty loading conditions. Derivations correspond to additional forces that can then be calculated.

The most accurate device in widespread use to date is the *force/torque (F/T) sensor*. It is a mechanical device that is capable of detecting one or more components of a six-dimensional wrench applied to it—for instance, a 3D force and a 3D torque (see section 4.1). Most commercially available F/T sensors use *strain gauges*. Simply put, a strain gauge is a metal (i.e., conductive) foil that changes its shape when a wrench is applied to it, and while doing so its electrical resistance changes. In a typical configuration (figure 7.3), an F/T sensor consists of an inner hub of solid metal that is suspended in an outer ring using three symmetrical, rectangular solid metal rods. Each metal rod is equipped with a strain gauge on each side (four per rod). Typically, sensors operate in pairs, one mounted orthogonal to

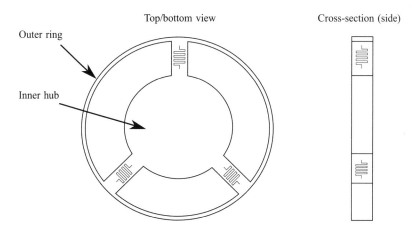

Figure 7.3
A force/torque sensor translates force and torque between two links of a robot via three metal rods that connect an inner hub to an outer ring. Here, one link of the robot connects to the outer ring and the other connects to the inner hub. Each metal rod is equipped with a strain gauge on each side, resulting in 12 sensors in total.

the other, resulting in a total of six sensor signals, from which we can compute forces and torques in three dimensions. Such F/T sensors are available as stand-alone components that can be mounted between an end-effector and a robot arm, or they are integrated within the robot's joints.

While accurate and precise, F/T sensors are plagued by a number of limitations: (1) high costs due to the high precision that is required during manufacturing; (2) size, with a standard F/T sensor usually as large as a human wrist; (3) low signal-to-noise ratio; (4) low bandwidth/responsivity; and (5) a single data point that is sparse in space and time. This last point becomes particularly clear when considering a robotic arm having multiple points of contact with an object. Here, a single sensor that measures forces and torques at the joint provides only very little information.

7.4.1 Measuring Pressure or Touch

To partially mitigate these limitations, roboticists have worked on a complementary capability: measuring the pressure applied on the robot's surface.

The skin is the largest human sense organ, and the human sense of touch is the oldest and the most important of our senses. To humans, contact and physical interaction are a resource rather than an impediment, and we are surprisingly proficient at leveraging touch in a variety of situations. Therefore, it is natural for roboticists to equip robots with similar capabilities in order to achieve performance levels comparable to that of humans.

A pressure sensor is a device that is capable of detecting either a contact/collision as binary data (in which case it is generally referred to as a touch sensor) or as a gradient of pressure applied to it. In general, the vertical pressure applied to the sensor is proportional to the one-dimensional vertical force that is applied to the direction normal to the sensor, and this makes a pressure sensor a good substitute for F/T sensors in specific applications (e.g., grasping). Additionally, pressure sensors are mostly based on measuring pressure through a change in *capacitance* rather than resistance (no different from the functioning of a touchscreen on a modern smartphone). When pressure is applied to a capacitor-like device (i.e., two conductive plates separated by an insulating material), the distance between the two plates reduces, causing a change in capacitance that can be readily measured.

As introduced in the context of series-elastic actuators, distance and force sensing are tightly related because of Hooke's law. Many touch and force sensors rely on light-based distance measurement (see section 7.5) in conjunction with a flexible material with known spring constant, such as using distance sensors to measure the deformation of an elastic dome from the inside (Youssefian, Rahbar, and Torres-Jara 2013) or measuring distance to objects through transparent rubber before touch and contact force after touch (Patel, Cox, and Correll 2018).

If compared to F/T sensors, pressure sensors provide a limited amount of information (one-dimensional versus six-dimensional), but they allow for high responsiveness; high density of sensing per square centimeter (up to tens of sensors per cm^2); low cost; and ease of miniaturization.

Human touch is not limited to pressure alone but also high-frequency information such as vibrations that are important when discerning different surfaces. Robots can replicate this capability by measuring vibrations on the order of hundreds of hertz by integrating accelerometers or microphones into a soft transducer and classifying spectral information (Hughes and Correll 2015).

In an extreme, it might be desirable to equip robots with an *artificial skin* that combines different sensing modalities for pressure, texture, temperature, or light, possibly also including cameras or actuators to change their appearance. While there exist a variety of commercial solutions, including pressurized double-layer skin that measures pressure differentials at select locations to detect contact, as well as capacitive solutions, robotic skins have not found widespread applications as of yet.

7.5 Sensors to Measure Distance

We have seen that there is a fluent transition from proprioceptive to exteroceptive sensors as measuring the robot's internal state is tightly related to its environment as soon as contact is made. In order to explore its environment from afar, measuring distance to individual objects has shown to be critical for the robot to navigate and identify obstacles and objects of interest.

The small form factor and low price of light-sensitive semiconductors have led to a proliferation of light-based sensors relying on a multitude of physical effects. These include reflection, phase shift, and time of flight. Other physical principles that are commonly used in distance sensors are radio (more commonly known as "radar") and sound.

7.5.1 Reflection

Reflection is one of the easiest and most immediate principles to take advantage of: The closer an object is, the more it reflects light, radio, or sound directed at it. We can then easily measure distance to objects that reflect the signal well and that are not too far away. In order to make these sensors as independent of an object's color as possible (but unfortunately not totally independent), infrared is most commonly chosen as the wavelength when using light. In contrast, sound will not be affected by a surface's color but by its surface properties and absorption characteristics.

A reflection-based distance sensor is made from two components: an emitter (that emits a signal such as infrared light) and a receiver (tasked with measuring the strength of the reflected signal). A typical response for an infrared distance sensor is shown in figure 7.4. The values obtained at an analog-digital converter correspond to the voltage at the infrared receiver and are saturated for low distances (flat line) and quadratically decrease afterward.

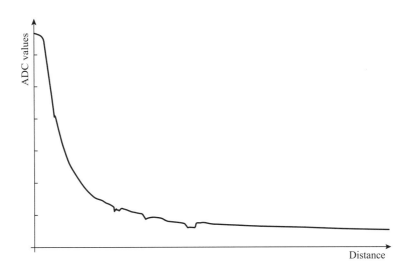

Figure 7.4
Real-world response of an infrared distance sensor as a function of distance. Units are left dimensionless intentionally.

7.5.2   Phase Shift

As shown in figure 7.4, reflection can only be precise if distances are short. Instead of measuring the strength (amplitude) of the reflected signal, laser distance sensors measure the phase difference of the reflected wave. In order to do this, the emitted light is modulated with a wavelength that exceeds the maximum distance the scanner can measure. If you were to use visible light and to do so at much slower speeds, you would see a light that keeps getting brighter, then gets darker, briefly turns off, and then starts getting brighter again.

Thus, if you were to plot the amplitude of the emitted signal over time (i.e., its brightness), you would see a wave that crosses zero when the light is dark. As light travels, this wave propagates through space with a constant distance (the wavelength) between its zero crossings. When it gets reflected, the same wave travels back (or at least parts of it get scattered right back). For example, modern laser scanners emit signals with a frequency of 5 megahertz (MHz), turning off 5 million times in one second. Together with the speed of light of approximately 300,000 kilometers per second, this leads to a wavelength of 60 meters (m) and makes such a laser scanner useful up to 30 m.

When the laser is now at a distance that corresponds exactly to one-half the wavelength, the reflected signal it measures will be dark at the exact same time its emitted wave goes through a zero-crossing. Going closer to the obstacle results in an offset that can be measured. Because the emitter knows the shape of the wave it emitted, it can calculate the phase difference between emitted and received signal. Knowing the wavelength it can now calculate the distance. As this process is mostly independent from ambient light, the estimates can be very precise.

As the laser distance measurement process is fast, such lasers can be combined with rotating mirrors to sweep larger areas, known as *laser range scanners* or *lidars*. Such systems have been combined into packages consisting of up to 64 scanning lasers and are nowadays vastly used in the autonomous driving space as they are capable of providing voluminous depth data of the enviornment around a car while driving. It is also possible to modulate projected images with a phase-changing signal, which is the operational principle of early "time of flight" cameras, which however is not an accurate description of their operation.

7.5.3   Time of Flight

The most precise distance measurement that light can provide is its time of flight. This can be done by counting the time a signal from an emitter becomes visible in a receiver. As light travels very fast ($3 \cdot 10^8$ m/s), this requires high-speed electronics that can measure time periods smaller than nanoseconds (*ns*) in order to achieve centimeter accuracy. In practice, this is done by combining the receiver with a very fast electronic shutter that operates at the same frequency of the emitted light. Because this timing is known, one can infer the time that light has traveled by measuring the quantity of photons that made it back from the reflective surface within one shutter period. As an example, light travels 15 meters (m) in 50 nanoseconds (ns). Therefore, it will take a pulse of 50 ns to return from an object at a distance of 7.5 m. If the emitter transmits a pulse of 50 ns length and then closes the receiver with a shutter, the receiver will receive more photons the closer the object is, but no photons if the object is farther than 7.5 m. Given a fast enough and precise circuit that acts as a shutter, it is sufficient to measure the actual amount of light that returns from the emitter.

Ultrasound distance sensors

Measuring the time of flight is considerably simpler when using sound waves to measure distance (sound travels at around 344 meters per second (m/s) in air). An ultrasound distance sensor operates by emitting an ultrasound pulse and measuring its reflection. Unlike a light-based sensor that measures the amplitude of the reflected signal, a sound-based sensor measures the time it took for the pulse to travel back and forth. This is possible because sound travels at much lower speed ($3 \cdot 10^2$ m/s) than light ($3 \cdot 10^8$ m/s). The fact that the sensor actually has to wait for the signal to return leads to a trade-off between range and bandwidth (see section 7.1): Allowing a longer range requires waiting longer for the signal to come back, which in turn limits how often the sensor can provide a measurement. Although ultrasound distance sensors have become progressively less common in robotics, they have an advantage over light-based sensors—namely, instead of sending out a ray, the ultrasound pulse results in a cone with an opening angle of 20 to 40 degrees. Because of this, such sensors are able to detect small obstacles without the requirement of directly hitting them with a ray. This property makes them the sensor of choice in specific applications, such as the automated parking assist technologies in modern cars.

7.6   Sensors to Sense Global Pose

So far, we have discussed sensors that allow the robot to measure the position of its own joints, its rotational velocity, its translational acceleration, forces from interaction with the environment, and distance to objects relative to its own pose. In order to reliably navigate in the environment, robots also need some notion of a world coordinate frame.

Localizing an object by triangulation goes back to ancient civilizations, where sailors oriented themselves using the stars. As stars are only visible during unclouded nights, seafarers have invented systems of artificial beacons emitting light, sound, and eventually radio waves. The most sophisticated of such systems is the Global Positioning System (GPS). GPS consists of satellites in orbit that are equipped with knowledge about their precise location and have synchronized clocks. These satellites broadcast a radio signal that travels at the speed of light and is coded with its time of emission. GPS receivers can therefore calculate the distance to each satellite by comparing time of emission and time of arrival. Because not only the position (x, y, z) but also the time difference between the GPS receiver's clock and the synchronized clocks of the satellites is unknown, four satellites are needed to obtain a "fix." Because of the way information from the satellites is coded, getting an initial fix can take on the order of minutes, but afterward it becomes available multiple times per second. GPS measurements are neither precise nor accurate enough for robotics applications and must be combined with other sensors, such as IMUs. (Note that the bearing shown on some GPS receivers is calculated from subsequent positions and is therefore meaningless if the robot is not moving.)

Various indoor GPS solutions exist that consist of either active or passive beacons that are mounted in the environment at known locations. Passive beacons—for example, infrared-reflecting stickers arranged in a certain pattern or 2D barcodes—can be detected using cameras, and their pose can be calculated from their known dimensions. Active beacons instead usually emit radio, ultrasound, or a combination thereof, which can then be used to estimate the robot's range to this beacon. In this domain, ultra-wideband radio in particular is common for relative localization indoors.

Take-Home Lessons

• Most of a robot's sensors either address the problem of determining the robot's pose or localizing and recognizing objects in its vicinity.

• Each sensor has advantages and drawbacks that are quantified in its range, precision, accuracy, and bandwidth. Therefore, robust solutions to a problem can only be achieved by combining multiple sensors with differing operation principles.

• Solid-state sensors (i.e., without mechanical parts) can be miniaturized and cheaply manufactured in quantity. As a result, a series of affordable IMUs and 3D depth sensors are available that will provide the data basis for localization and object recognition on mass-market robotic systems.

Exercises

1. Given a laser scanner with an angular resolution of 0.01 rad and a maximum range of 5.6 meters, what is the minimum range d a robot needs to be from an object of 1 centimeter width to definitely sense it (i.e., hit it with at least one of its rays)? You can approximate the distance between two rays with the arc length.

2. Why does the bandwidth of an ultrasound-based distance sensor decrease significantly when increasing its dynamic range, but that of a laser range scanner does not for typical operation?

3. You are designing an autonomous electric car to transport goods on campus. Since you are worried about cost, you are thinking about whether to use a laser scanner or an ultrasound sensor for detecting obstacles. As you drive rather slow, you are required to sense up to 15 meters. The laser scanner you are considering can sense up to this range and has a bandwidth of 10 Hz. Assume 300m/s for the speed of sound.

a) Calculate the time it takes until you hear back from the ultrasound sensor when detecting an obstacle 15m away. Assume that the robot is not moving at this point.

b) Calculate the time it takes until you hear back from the laser scanner. (Hint: You don't need the speed of light for this, because the answer is in the specs above.)

c) Assume now that you are moving toward the obstacle. Which sensor will give you a measurement that is closer to your real distance at the time of reading, and why?

4. Pick an educational robot platform of your choice and make a list of its sensors.

5. Construct a simple range scanner by mounting an ultrasound sensor onto a servo motor. Implement a scanning routine that allows you to collect the raw data and display it on the screen. Can you see simple features such as corners and openings?

6. Explore the internet for do-it-yourself robotics shops. What kind of sensors do they offer? What are the interfaces these sensors provide?

7. Pick a physical sensor that you have access to. Can you design an experiment to characterize its precision and accuracy?

8. Your task is to design a sensor that can detect the remaining void in a parcel for an e-commerce application.

a) What sensors do you think would allow you to measure the volume of the content in the box?

b) What additional sensors could you use, assuming the box is moving on a conveyor belt?

c) What additional information would you need to know in order to differentiate between box content, the box itself, and the environment around the box? What sensors could you use to get this information?

d) Additional sensors are not within your budget. What kind of measures could you take to reduce the amount of information required?

9. Your task is to design an autonomous cart that can automatically dock with shelves in the environment.

a) What kind of sensors could you use to locate the shelf in the environment? Assume that the shelf is the only object in a certain target area.

b) What kind of physical measures could you take to simplify detection of the shelf?

c) What kind of sensors could you use to detect whether the shelf is in a suitable position for docking?

d) What kind of physical measures could you take to simplify the sensing process?

10. You are designing a competitive controller for the "Ratslife" game. What kind of information does the environment provide, and what kind of sensor would you need to exploit it?

III COMPUTATION

8 Vision

Vision is one of the most information-rich sensor systems both humans and robots have available. However, efficiently and accurately processing the wealth of information that is generated by vision sensors is still a key challenge in the field. The goals of this chapter are to:

- Introduce the concept of images as two-dimensional signals.
- Provide an intuition of the wealth of information hidden in low-level information.
- Introduce basic convolution and threshold-based image processing algorithms.

8.1 Images as Two-Dimensional Signals

Images are captured by cameras containing matrices of charge-coupled devices (CCD) or similar semiconductors such as complementary metal–oxide semiconductors (CMOS) that can turn photons into electrical signals. These matrices can be read out pixel by pixel and turned into digital values—for example, an array of 640 by 480 three-byte tuples corresponding to the red, green, and blue (RGB) components the camera has seen. An example of such data is shown in figure 8.1; for simplicity, we show only one color channel. Looking at the data in the matrix clearly reveals the white tile within the black tiles at the lower-right corner of the chessboard. Higher values correspond to brighter colors (white) and lower values to darker colors. We also observe that although the tiles have to have the same color, the actual values differ quite a bit. It might make sense to think about these values much like we would if the data were a one-dimensional (1D) signal: Taking the "derivative" along the horizontal rows, for example, would indicate areas of big changes, whereas a "frequency" histogram of an image would indicate how quickly values change. Areas with smooth gradients (e.g., black and white tiles) would then have low frequencies, whereas areas with strong gradients would contain high-frequency information.

This language opens the door to a series of signal processing concepts, such as low-pass filters (suppressing high-frequency information), high-pass filters (suppressing low-frequency information), or band-pass filters (letting only a range of frequencies pass);

ISS017E011574

Figure 8.1
A chessboard floating inside the International Space Station (ISS) with astronaut Gregory Chamitoff. The inset shows a sample of the actual data recorded by the image sensor. One can clearly recognize the contours of the white tile. *Source*: NASA

analysis of the frequency spectrum of the image (the distribution of content at different frequencies); or "convolving" the image with another two-dimensional function. The next sections provide both an intuition of what kind of meaningful information is hidden in such abstract data and concrete examples of signal processing techniques that make this information appear.

8.2   From Signals to Information

Unfortunately, many phenomena that often have very different or even opposite meaning look very similar when looking at the low-level signal. For example, drastic changes in color values do not necessarily mean that the color of a surface indeed has changed. Similar patterns are generated by depth discontinuities, specular highlights, changing lighting conditions, or surface orientation changes. These phenomena—some of which are illustrated in figure 8.2—make computer vision a hard problem.

This example illustrates that signals and data alone are not sufficient to understand a phenomenon but require context. Here, the context does not only refer to surrounding signals but also high-level conceptional knowledge such as the fact that light sources create shadows and specular highlights, that objects in the front appear larger, and so on. The importance of such conceptional knowledge is illustrated in figure 8.3. Both images show an identical landscape, but in one picture the landscape appears to be speckled with craters while in the other picture it has bubble-like hills. At first glance, both scenes are illuminated

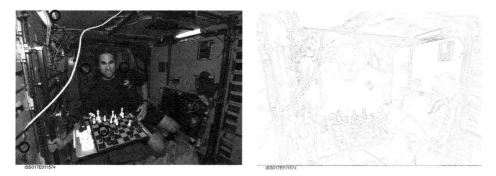

Figure 8.2
Inside the ISS. The left side shows circled areas in which pixel values change drastically (right side). Underlying effects that produce similar responses are change in surface properties (1), depth discontinuities (2), specular highlights (3), changing lighting conditions such as shadows (4), or surface orientation changes (5).

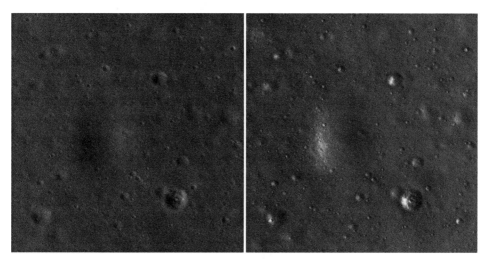

Figure 8.3
Pictures of the Apollo 15 landing site during different times of the day. The landscape is identical but appears to be either speckled with craters (left) or hills (right). Knowing that the sun is illuminating the scene from the left and right, respectively, explains this effect. *Source*: NASA/Goddard Space Flight Center/Arizona State University.

from the left, suggesting a change in the landscape. However, once information that the sun is illuminating one picture from the left and the other from the right is available, the paradox becomes clear: the variable illumination makes the craters look like bumps under come conditions.

More surprisingly, conceptual knowledge is often sufficient to make up for the lack of low-level cues in an image. An example is shown in figure 8.4. Here, a Dalmatian dog can

Figure 8.4
The image of a Dalmatian can be clearly recognized by most spectators even though low-level cues such as edges are only present for ears, chin, and parts of the dog's legs. The contours of the animal are highlighted in a flipped version of the image in the inset.

be clearly recognized despite the absence of cues for its outline (i.e., by simply extrapolating its appearance and pose from conceptual knowledge).

These examples illustrate both the advantages and drawbacks of a signal processing approach to computer vision. While an algorithm will detect interesting signals even where we don't see or expect them (because of conceptional bias), image understanding not only requires low-level processing but also intelligent combination of the spatial relationship between low-level cues and conceptual knowledge about the world. As we will later see (chapter 10), this can be accomplished through convolutional neural networks that provide a single pipeline to process information at different scales—ranging from extracting local features to examining their spatial relationships with each other.

8.3   Basic Image Operations

Basic image operations can be thought of as a filter that operates in the frequency domain or in the space (intensity/color) domain. Although most filters directly operate in the intensity domain, knowing how they affect the frequency domain is helpful in understanding the filter's function. For example, a filter that is supposed to highlight edges such as the one shown in figure 8.2 should suppress low frequencies (i.e., areas in which the color values do not change much) and amplify high-frequency information (i.e., areas in which the color values change quickly). The goal of this section is to provide a basic understanding of how basic image processing operation works. It is important to note that the methods presented here, while still valid, have been superseded by more sophisticated implementations that are widely available as software packages or within desktop graphic software.

8.3.1   Threshold-Based Operations

In order to find objects with a certain color or edge amplitude, thresholding an image by running a Boolean operation over each pixel (e.g., true for pure green pixels and false otherwise) will lead to a binary image that contains "true-false" regions that fit the desired criteria. Thresholds make use of operators like $>, <, \leq, \geq$, and combinations thereof. There are also adaptive versions that adapt/update the thresholds locally to make up for changing lighting conditions, for example. Although thresholding is simple if compared to other techniques introduced later in this chapter, finding correct threshold values is a hard problem. In particular, actual pixel values change drastically when lighting conditions change, and there is no such thing as "red" or "green" when inspecting the actual values under different conditions.

8.3.2   Convolution-Based Filters

A filter can be implemented using the *convolution* operator $\star$ that convolves function $f()$ with function $g()$:

$$f(x) \star g(x) = \int_{-\infty}^{\infty} f(\tau)g(x-\tau)d\tau, \tag{8.1}$$

where $g()$ is defined as *filter*. The convolution essentially "shifts" the function $g()$ across the function $f()$ while multiplying the two (see the video linked via QR code in this section). As images are discrete signals, the convolution is consequently discrete:

$$f[x] \star g[x] = \sum_{i=-\infty}^{\infty} f[i]g[x-i]. \tag{8.2}$$

Additionally, given that images are two-dimensional signals, the convolution is two-dimensional as well:

$$f[x,y] \star g[x,y] = \sum_{i=-\infty}^{\infty} \sum_{j=-\infty}^{\infty} f[i,j]g[x-i,y-j]. \tag{8.3}$$

Although we have defined the convolution from negative infinity to infinity, both images and filters are usually finite. Images are constrained by their resolution, and filters are usually much smaller than the images themselves. Also, the convolution is commutative, therefore equation (8.3) is equivalent to

$$f[x,y] \star g[x,y] = \sum_{i=-\infty}^{\infty} \sum_{j=-\infty}^{\infty} f[x-i, y-j]g[i,j]. \tag{8.4}$$

Gaussian smoothing

One of the most basic (and important) filters is the *Gaussian filter*. The Gaussian filter is shaped like the Gaussian bell function and can be easily stored in a two-dimensional matrix. Implementing a Gaussian filter is surprisingly simple. For example:

$$g(x,y) = \frac{1}{10} \cdot \begin{pmatrix} 1 & 1 & 1 \\ 1 & 2 & 1 \\ 1 & 1 & 1 \end{pmatrix}. \tag{8.5}$$

Using this filter in equation 8.4 on an infinitely large image $f()$ leads to

$$f[x,y] \star g[x,y] = \sum_{i=-1}^{1} \sum_{j=-1}^{1} f[x-i, y-j]g[i,j], \tag{8.6}$$

assuming that $g(0,0)$ is the center of the matrix. What now happens is that each pixel $f(x,y)$ becomes the average of that of its neighbors, with its previous value weighted twice that of its neighbors (because $g(0,0) = 0.2$). More concretely:

$$
\begin{aligned}
f(x,y) = \quad & uf(x+1,y+1)g(-1,-1) + f(x+1,y)g(-1,0) + f(x+1,y-1)g(-1,1) \\
& + f(x,y+1)g(0,-1) \quad + f(x,y)g(0,0) \quad + f(x,y-1)g(0,1) \\
& + f(x-1,y+1)g(1,-1) + f(x-1,y)g(1,0) \quad + f(x-1,y-1)g(1,1).
\end{aligned}
\tag{8.7}
$$

Doing this for all x and all y is equivalent to physically "sliding" the filter $g()$ along the image.

An example of the Gaussian filter in action is shown in figure 8.5. The filter acts as a *low-pass filter*, suppressing high-frequency components. Indeed, noise in the image is suppressed, leading also to a smoother edge image, which is shown to the right.

Edge detection

Edge detection can be achieved using another convolution-based filter, the *Sobel* kernel:

$$s_x(x,y) = \begin{pmatrix} -1 & 0 & 1 \\ -2 & 0 & 2 \\ -1 & 0 & 1 \end{pmatrix} \qquad s_y(x,y) = \begin{pmatrix} 1 & 2 & 1 \\ 0 & 0 & 0 \\ -1 & -2 & -1 \end{pmatrix}. \tag{8.8}$$

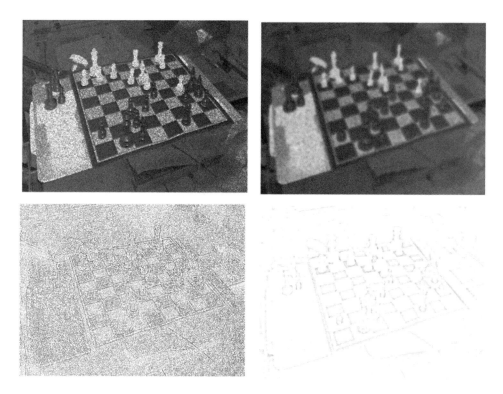

Figure 8.5
A noisy image before (top left) and after filtering with a Gaussian kernel (top right). Corresponding edge images are shown underneath.

Here, $s_x(x, y)$ can be used to detect vertical edges, whereas $s_y(x, y)$ highlights horizontal edges. Edge detectors such as the *Canny* edge detector therefore run at least two such filters over an image to detect both horizontal and vertical edges.

Difference of Gaussians

An alternative method for detecting edges is the *difference of Gaussians* (DoG) method. The idea is to subtract two images that have each been filtered using a Gaussian kernel with different width. Both filters suppress high-frequency information, and their difference therefore leads to a *band-pass* filtered signal, from which both low and high frequencies have been removed. As such, a DoG filter acts as a capable edge detection algorithm. Here, one kernel is usually four to five times wider than the other, therefore acting as a much stronger filter.

Differences of Gaussians can also be used to approximate the *Laplacian of Gaussian* (LoG), which is the sum of the second derivatives of a Gaussian kernel. Here, one kernel is

Image I   Erosion I⊖B   Dilatation I⊕B   Opening IoB = (I⊖B)⊕B

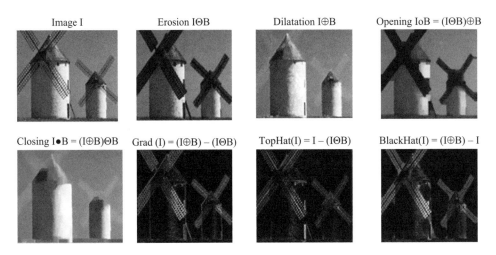

Closing I●B = (I⊕B)⊖B   Grad (I) = (I⊕B) – (I⊖B)   TopHat(I) = I – (I⊖B)   BlackHat(I) = (I⊕B) – I

Figure 8.6
Examples of morphological operators erosion and dilation and combinations thereof. *Source*: OpenCV documentation, BSD.

roughly 1.6 times wider than the other. The band-pass characteristic of DoG and LoGs are important as they highlight high-frequency information such as edges yet suppress high-frequency noise in the image.

8.3.3 Morphological Operations

Another class of filters are morphological operators that consist of a kernel describing the structure of the operation (this can be as simple as an identity matrix) and a rule on how to change a pixel value based on the values in the neighborhood defined by the kernel. Important morphological operators are *erosion* and *dilation*. The erosion operator assigns a pixel a value with the minimum value that it can find in the neighborhood defined by the kernel. The dilation operator assigns a pixel a value with the maximum value it can find in the neighborhood defined by the kernel. This is useful to fill holes in a line or remove noise, for example. A dilation followed by an erosion is known as a "closing," and an erosion followed by a dilation as an "opening." Subtracting eroded and dilated images from each other can also serve as an edge detector. Examples of such operators are shown in figure 8.6.

8.4 Extracting Structure from Vision

A remarkable property of vision is the ability to provide both semantic (*qualities* of the scene, such as what is in it) and metric (*quantities* of the scene, such as sizes and distances) information. Extracting semantic information is nowadays heavily reliant on

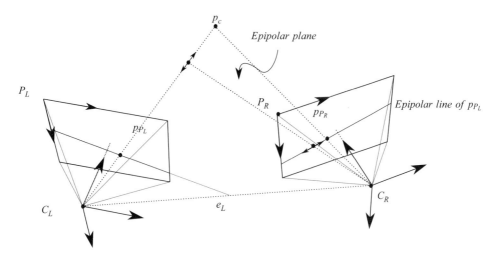

Figure 8.7
Schematic of correlating features across images, used in order to extract three-dimensional information from two-dimensional views.

machine learning, which is explained at a high level in section 8.5. The extraction of metric information, however, can be accomplished by leveraging geometric relationships, which we will describe here.

Figure 8.7 shows a high-level schematic of the relationships between an image frame and another—both observing the same point. Here, we do not draw a distinction between these two frames being either spatially or temporally correlated, which are two distinct problems in robotics: In *stereo vision*, two cameras are rigidly attached to one another (spatial correlation) and are acquiring images of the same scene; in *structure from motion*, a single camera is moved through a scene and a pair of images from the single camera are related to one another using a transform matrix (temporal correlation). In either case, it is possible to identify the "projection center" of the camera frames as C_L and C_R; they are related to one another through T_{LR}, which is defined as the transform matrix from the left to the right frame. In stereo vision, this transform is known as the *sensor extrinsics*, a six degrees of freedom (6-DoF) quantity that must be estimated through calibration. In structure from motion, this transform quantifies the motion applied to the camera, which can be estimated through localization (see chapter 16).

Note that since the camera pair takes two images of the same scene, two projections into the corresponding image planes of the same point in the world can be correlated with one another to determine the point's 3D position. This may be accomplished naively by identifying the point p_{P_L} in C_L and searching for that point p_{P_R} in C_R. Crucially, the nature of projections of 3D points into the camera frame is a known operation and, in fact, a very simple one. In particular, a 3D point in the world can be projected into the camera frame

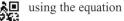

using the equation

$$\begin{pmatrix} u \\ v \\ 1 \end{pmatrix} = KT \begin{pmatrix} x \\ y \\ z \\ 1 \end{pmatrix},$$ (8.9)

where K is known as the *camera intrinsic matrix* and T is the matrix form of the transform between the camera and some global coordinates in which the point (x, y, z) is expressed. Note that the camera intrinsic matrix is another *calibrated* quantity, instantiated by two optical center parameters and two scaling parameters. Importantly, it is possible for two projected points representing a single point in three-dimensional (3D) space to be calculated directly through triangulation on these two-dimensional (2D) point pairs in images alone. Using the same math as in equation (8.9) but expressing C_L as the global coordinate system, we can relate the 3D coordinate of the point to the two 2D measurements, camera intrinsic matrix, and T_{LR}:

$$\begin{pmatrix} u_R \\ v_R \\ 1 \end{pmatrix} = KT_{LR} \begin{pmatrix} x \\ y \\ z \\ 1 \end{pmatrix} = KT_{LR}K^{-1} \begin{pmatrix} u_L \\ v_L \\ 1 \end{pmatrix}.$$ (8.10)

Note that this expression is frequently given in terms of what is known as the "essential matrix," which is nominally a technique to solve this problem for uncalibrated cameras. This expression, clearly induced by the geometry expressed in figure 8.7, allows for alternatively solving the values making up the entries of the essential matrix and not those of the camera intrinsic and extrinsic parameters.

Looking back at the geometry in figure 8.7, it may be noted that p_{P_L} lies on a line that extends from the center of camera C_L to the point p. However, there is ambiguity of the depth along the ray that is cast from C_L to p_c $\overrightarrow{C_L p_c}$. In order to disambiguate this depth, the line between p_{P_L} and the center of projection of P_L, which is known as the "epipolar line," may be projected into P_R, which creates the so-called epipolar line of p_L in C_R. It is along this line that the point p_c projected into P_R may be found. Most notably, therefore, the search for the point p_{P_R} may be reduced to a line search along the projected epipolar line. This line search is much more efficient than finding the projection of point p_c in the whole image plane of P_R, and therefore it allows for rapidly sped-up geometry calculations across image pairs.

Extracting metric information from images requires us to uniquely identifying identical points in each image. A simple solution to this problem is what is known as *structured light*, which is illustrated in figure 8.8. Thanks to the continuously increasing efficiency of computational systems, a lightweight version of such an approach that could be implemented at small scale and low cost emerged around 2010, establishing a novel standard in robotic sensing.

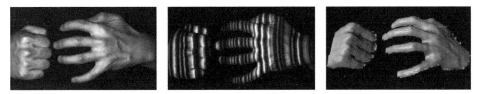

Figure 8.8
From left to right: two complex physical objects, a pattern of straight lines, typically projected in a pattern of colors to facilitate differentiating them, and their deformation when hitting the surfaces, reconstructed 3D shape. *Source*: Zhang, Curless, and Seitz 2002, 1.

Instead of using line patterns, infrared-based depth image sensors use a speckle pattern (a collection of randomly distributed dots with varying distances). Identifying identical points in two images simply requires one to search for blobs with similar size that are close to each other.

8.5   Computer Vision and Machine Learning

The algorithms described here still form the basis of most image understanding pipelines and make feature detection (chapter 9) tractable. With the advent of so-called convolutional neural networks (chapter 10), basic signal processing such as described here is now often wrapped into the image understanding problem. While this makes it less important to implement such algorithms oneself, understanding what convolution, morphological operations, and thresholds do to visual information remains still relevant to meaningfully compose neural networks and make them less of a black box.

Take-Home Lessons

1. Unlike the sensors from chapter 7, our brains can directly process the 2D information that is captured by a vision sensor. It is difficult to unthink the amount of processing that we perform automatically, augmenting the signal with knowledge and other information that the computer does not necessarily have.

2. Algorithms described in this chapter aim at reducing information to a lower-dimensional space by removing noise and other spurious information, making the related challenge of understanding the data more tractable.

3. There is a trade-off between making the data stream more tractable and preserving actual information. As computers and algorithms, in particular machine learning, become more powerful, modern vision systems often blend preprocessing and actual image understanding into a single pipeline.

Exercises

1. Below are shown multiple "kernels" that can be used for convolution-based image filtering.

1	1	1
1	2	1
1	1	1

0	−1	0
0	−1	0
0	−1	0

1	1	1
1	−4	1
1	1	1

a) Identify the kernel that can blur an image.

b) What kind of features can be detected by the other two kernels?

2. How many for-loops are needed to implement a 2D convolution? Explain your reasoning.

3. Use an appropriate robot simulation environment that allows you to access a simulated camera in a world with simple features such as geometric shapes of different color.

a) Implement a thresholding algorithm that allows you to black out anything but an object of a specific color. Is a simple threshold enough? Why not? Can you black out an object using a low and and a high threshold?

b) Implement a smoothing algorithm by performing both a convolution with a Gaussian kernel as well as a series of morphological operations. Experiment with kernels of different width and different steepness. What are the advantages and drawbacks of using morphological operations over a simple Gaussian filter?

c) Implement an edge detection algorithm—for example, by performing a convolution with a Sobel kernel. Experiment with different kernels. What else do you need to do to create an image that only contains edges?

4. Can you think about a smoothing algorithm that will only smoothen small amounts of noise but maintain edges? What kind of filtering algorithms could you combine to achieve this goal?

5. Explore the internet for a computer vision toolbox that supports your language of choice. What do you find? Does the toolbox implement all of the algorithms in this chapter? Solve the above assignments using the toolbox's built-in functions.

6. Use an appropriate robot simulation environment that allows you to simulate two cameras that are at a known distance in the same plane. Use simple geometric objects such as a red ball and compute their distance using stereo disparity.

9 Feature Extraction

A robot can obtain information about its environment by both active (e.g., ultrasound, light, and laser) or passive sensing (e.g., acceleration, magnetic field, or cameras). In only limited cases is this information directly useful to a robot without significant preprocessing. For example, before being able to arrive at semantic information such as "I'm in the kitchen," "this is a cup," or "this is a horse," one must first identify higher-level *features* and correlate these features with the information of interest.

The goal of this chapter is to introduce the notion of features and understand standard feature detectors, including the following:

- The Hough transform to detect lines, circles, and other shapes
- Numerical methods such as least squares, split-and-merge, and random sample and consensus (RANSAC) to find high-level features in noisy data
- Scale-invariant features (SIFT)

9.1 Feature Detection as an Information-Reduction Problem

The information generated by sensors can be quite voluminous. For example, a simple webcam generates 640×480 color pixels (red, green, and blue) or 921,600 bytes around 30 times per second. A single-ray laser scanner provides around 600 distance measurements 10 times per second in the form of a point cloud. Consider for a moment the information that a robot *requires* in order to solve its problems, however. The volume of data resulting from most sensors seems much greater than the amount of information required to answer the query, "What are the dimensions of this room?" Consider, for example, the maze-solving competition "Ratslife" (section 1.3) in which the robot's camera can be used to recognize one of 48 different patterns (figure 1.3) that are distributed in the environment, or the presence or absence of a charger, essentially reducing hundreds of bytes of camera data to 6 bits of content ($2^6 = 64$ different values). The goal of most image processing algorithms is therefore to first reduce information content in a meaningful way and then extract relevant information. In chapter 8, we were introduced to convolution-based filters such as blurring, detecting edges, or binary operations such as thresholding. We are now interested in methods to extract higher-level features such as lines and techniques using these types of processing algorithms.

9.2   Features

Lines are particularly useful features for localization and can correspond to walls in two-dimensional (2D) laser scans, edges in three-dimensional (3D) laser scans, markers on the floor, or corners detected in a camera image. Whereas a Sobel filter (section 8.3.2) can help us to highlight lines and edges in images, additional algorithms are needed to identify each line and extract structured information such as positions and orientations. This structured information can then aid in identifying these lines and edges through multiple observations, which allows us to identify persistent structures and reason over both their and our pose as we move around or past them.

A desirable property of a feature is that its extraction is repeatable and robust to rotation, scale, and noise in the data. We need feature detectors that can extract the same feature from sensor data, even if the robot has slightly turned or moved farther or closer to the feature. Ideally the same feature could also be extracted if there is some noise affecting the sensor. There are many feature detectors available that accomplish this. Prominent examples are the *Harris corner detector*, which detects points in the image where vertical and horizontal lines cross, and the scale-invariant feature transform (SIFT) detector, which identifies features through maxima in the difference of Gaussian image (section 8.3.2) at various spatial scales. Feature detection is important far beyond robotics and is, for example, used in hand held cameras that can automatically stitch images together and image indexing on the internet. In image stitching, feature detectors will "fire" (identify a feature) on the same features in two images taken from slightly different perspectives; these matched features provide a geometric template between the images where information is shared, thereby allowing for the two images to be concatenated.

This chapter focuses on two important classes of features in images: line features and SIFT. Both classes of features provide tangible examples for the least squares and RANSAC algorithms, which may be used to harness feature information to solve problems and are also introduced in this chapter. Together, these classes of features are fairly representative of all features used in robotics, and they have been chosen for their simplicity, providing a basis for understanding the function of more complex feature detectors.

These hand-coded feature detectors are in contrast to entirely self-learned feature detectors based on deep neural networks, which are treated in chapter 10. Although neural network–based methods often outperform hand-coded features, hand-coded features remain relevant for environments in which learning is unfeasible, to preprocess data before subjecting it to a learning-based method or to understand what kind of architecture is required to solve a specific goal.

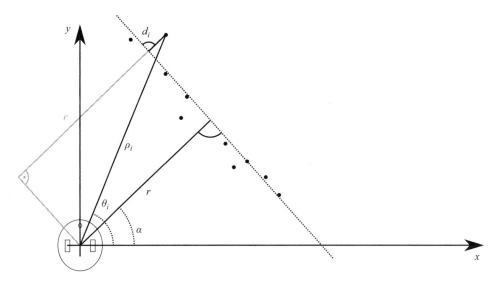

Figure 9.1
A 2D point cloud recorded by a laser scanner or similar device. A line (dashed) is fitted through the points in a least-squares sense.

9.3   Line Recognition

Why are lines a useful feature? As you will see in part IV of this book, on "uncertainty," the key challenge in estimating a robot's pose is unreliable odometry, in particular when it comes to turning. Here, a simple infrared sensor measuring the distance to a wall can provide the robot with a much better sense for what actually happened during the turn. Similarly, if a robot has the ability to track markers in the environment using vision, it gets another measurement on how much the robot is actually moving. How information from odometry and other sensors can be fused not only to localize the robot but also to create maps of its environment will be a significant focus in the remainder of this book.

A laser scanner or similar device pointed at a wall will return a measurement of N points at position (x_i, y_i) in the robot's coordinate system. These points can also be represented in polar coordinates (ρ_i, θ_i). We can now imagine a line running through these points that is parameterized with a distance r and an angle α. Here, r is the distance of the robot to the wall and α its angle. As all sensors are noisy, each point will have distance d_i from the "optimal" line running through the points. These relationships are illustrated in figure 9.1.

9.3.1   Line Fitting Using Least Squares

Using simple trigonometry we can now write

$$\rho_i \cos(\theta_i - \alpha) - r = d_i. \tag{9.1}$$

Different line candidates—parameterized by r and α—will have different values for d_i. We can now write an expression for the total error $S_{r,\alpha}$ as

$$S_{r,\alpha} = \sum_{i=1}^{N} d_i^2 = \sum_i (\rho_i \cos(\theta_i - \alpha) - r)^2. \qquad (9.2)$$

Here, we square each individual error to account for the fact that a negative error (i.e., a point left of the line) is as bad as a positive error (i.e., a point right of the optimal line). In order to optimize $S_{r,\alpha}$, we need to take the partial derivatives with respect to r and α and set them to zero, indicating the functions are at their minimum or maximum:

$$\frac{\partial S}{\partial \alpha} = 0 \qquad \frac{\partial S}{\partial r} = 0. \qquad (9.3)$$

Then we can solve the results of equation (9.3) for r and α. Here, the symbol ∂ indicates that we are taking a partial derivative. Solving for r and α is algebraically possible (Siegwart, Nourbakhsh, and Scaramuzza 2011):

$$\alpha = \frac{1}{2} \text{atan} \left(\frac{\frac{1}{N} \sum \rho_i^2 \sin 2\theta_i - \frac{2}{N^2} \sum \sum \rho_i \rho_j \cos \theta_i \sin \theta_j}{\frac{1}{N} \sum \rho_i^2 \cos 2\theta_i - \frac{1}{N^2} \sum \sum \rho_i \rho_j \cos(\theta_i + \theta_j)} \right), \qquad (9.4)$$

and

$$r = \frac{\sum \rho_i \cos(\theta_i - \alpha)}{N}. \qquad (9.5)$$

Therefore, using our proximity sensors, we can calculate the distance and orientation of a wall relative to the robot's positions, or the height and orientation of a line in an image, based on a collection of points that we believe might belong to a line.

This approach is known as the *least-squares method* and can be used to fit data to any parametric model (i.e., a model that has numbers sought to make it fit our data best). The general approach is to describe the fit between the data and the model in terms of a difference, known as an "error." The best fit will minimize this error—that is, the error will have a zero derivative for the best parameters. If the result cannot be obtained analytically as in this example, numerical methods have to be used to find the best fit that minimizes the error.

9.3.2   Split-and-Merge Algorithm

It is often unclear how many lines there are and where a line starts and ends. This creates a challenge for the matching-and-estimation strategy discussed previously. Looking through the camera, for example, we will see vertical lines corresponding to wall corners and horizontal lines that correspond to wall-floor intersections and the horizon; using a distance sensor, the robot might detect a corner. We therefore need an algorithm that can separate point clouds into multiple lines. One possible approach is to find the outlier with the

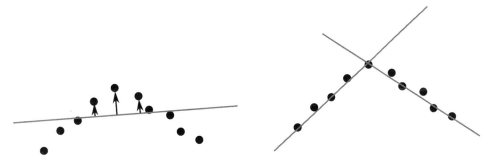

Figure 9.2
Split-and-merge algorithm. Initial least-squares fit of a line is shown on the left. Splitting the dataset at the point with the highest error (after picking a direction) allows fitting two lines with overall lesser error.

strongest deviation from a fitted line and split the line at this point, as illustrated in figure 9.2. This can be done iteratively until each line has no outliers above a certain threshold.

9.3.3   RANSAC: Random Sample and Consensus

If the number of outliers is large, a least-squares fit will generate poor results because it will be the "best" fit that accomodates both inliers and outliers. Note that this generally results in a very poor fit, since the least-squares fit mathematically assigns a significant weight to outliers—not treating them as a single measurement to be discarded but a cumulative error to be reduced in balance with those of the inliers. In this way, split-and-merge algorithms will fail as they are extremely sensitive to outliers: depending on the actual parameters, every outlier will split a potential line into two.

A powerful solution to this problem is to randomly sample possible lines and keep those that satisfy a certain desired quality, given by the number of points being somewhat close to the best fit. This is illustrated in figure 9.3, with darker lines corresponding to better fits. RANSAC usually requires two parameters—namely, the number of points required to consider a line to be a valid fit and the maximum d_i from a line to consider a point an inlier and not an outlier. The algorithm proceeds as follows: Select two random points from the set and connect them with a line. Grow this line by d_i in both directions and count the number of inliers. Repeat until one or more lines that have a sufficient number of inliers are found or a maximum number of iterations is reached. RANSAC is applied frequently when it comes to feature detection and matching because it provides a systematic routine for separating inliers from outliers even in very noisy data.

The RANSAC algorithm is fairly easy to understand in the line-fitting application but can be used to fit arbitrary parametric models to any-dimensional data. Here, its main strength is to cope with noisy data.

Given that RANSAC is random, finding a really good fit can be computationally intensive and time-consuming. Therefore, RANSAC is usually used only as a first step to get an

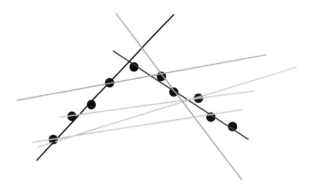

Figure 9.3
RANSAC. Random lines are evaluated by counting the number of points close by ("inliers"); darker lines are better fits.

initial estimate, which can then be improved by some kind of local optimization, such as least squares.

9.3.4   The Hough Transform

The Hough transform can best be understood as a voting scheme to guess the parameterization of a feature such as a line, circle, or other curve (Duda and Hart 1972). For example, a line might be represented by $y = mx + c$, where m and c are the gradient and offset. A point in this parameter space (or "Hough-space") then corresponds to a specific line in x-y-space (or "image-space"). The Hough transform now proceeds as follows: For every pixel in the image that could be part of a line (e.g., white pixels in a thresholded image after Sobel filtering), construct all possible lines that intersect this point. (The image drawn would look like a star.) Each of these lines has a specifc m and c associated with it, for which we can add a white dot in Hough-space. Continuing to do this for every pixel of a line in an image will yield many $m - c$ pairs, but only one that is common among all those pixels of the line in the image: the actual $m - c$ parameters of this line. Thinking about the number of times a point was highlighted in Hough-space as brightness will turn a line in image-space into a bright spot in Hough-space (and the other way around). In practice, a polar representation is chosen for lines, as shown in figure 9.4. The Hough transform also generalizes to other parameterizations such as circles.

9.4   Scale-Invariant Feature Transforms

SIFT algorithms are a class of techniques that allow one to extract features that are easily detectable across different scales (or distances to an object), independent of their rotation, and to some extent robust to perspective transformations and illumination changes. An early SIFT algorithm (Lowe 1999), which has lost some popularity because of its

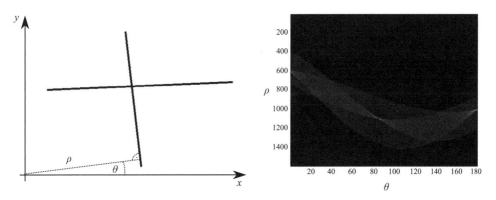

Figure 9.4
Lines in an image (left) transposed into Hough-space ρ (distance from origin) and θ (angle of normal with respect to origin). Bright spots in the Hough image (right) correspond to parameters that have received the most "votes" and clearly show the two lines at around 90 and 180 degrees.

licensing requirements, has been replaced with SURF (speeded-up robust feature) (Bay, Tuytelaars, and Van Gool 2006) and ORB (Rublee et al. 2011), both of which are freely available. As the arithmetic behind SURF is slightly more involved, we focus on the intuition behind SIFT and encourage the reader to download and play with the various open-source implementations of other feature detectors that are freely available.

9.4.1   Overview

SIFT proceeds in multiple steps. Descriptions of the algorithm often include its application to object recognition, but these algorithms are independent of the feature generation step.

1. Use the difference of Gaussians (DoG) method at different scales:

a) Generate multiple scaled versions of the same image by re-sampling every second, fourth, and so on pixel (up to the desired scale).

b) Filter each scaled picture with various Gaussian filters of different variance.

c) Calculate the difference between pairs of filtered images. This is equivalent to a DoG filter.

2. Detect local minima and maxima in the DoG images across different scales (figure 9.5, left) and reject those with low contrast (figure 9.5, right).

3. Reject extrema that are along edges by looking at the second partial derivatives in image-space around each extremum (figure 9.5, right). Edges have a much larger principal curvature across them than along them.

4. Assign a "magnitude" and "orientation" to each remaining extremum, now called a "key-point." The magnitude is the squared difference between the DoG filter response at the present pixel and the neighboring pixels. The orientation is the arctangent between the DoG

Figure 9.5
After scale space extrema are detected (left), the SIFT algorithm discards low contrast keypoints (center) and then filters out those located on edges (right). *Source*: Lukas Mach CC-BY 3.0.

differences in the y direction and the x direction. These calculations are made for all pixels in a fixed neighborhood around the initial keypoint (e.g., in a 16×16 pixel neighborhood).

5. Collect orientations of neighboring pixels in a histogram (e.g., 36 bins, each covering 10 degrees). Maintain the orientation corresponding to the strongest peak and associate it with the keypoint.

6. Repeat step 4, but for four 4×4 pixel areas around the keypoint in the image scale that has the most extreme minima/maxima. Here, only 8 bins are used for the orientation histogram. As there are 16 histograms in a 16×16 pixel area, the feature descriptor has 128 dimensions.

7. Normalize, threshold, and again normalize the feature descriptor vector to make it more robust against illumination changes.

8. Group local gradient magnitude and orientation into bins and create a 128-dimensional feature descriptor.

The resulting 128-dimensional feature vectors are now scale-invariant (because of step 2), rotation-invariant (result of step 5), and robust to illumination changes (result of step 7).

9.4.2   Object Recognition Using Scale-Invariant Features

Scale-invariant features of training images can be stored in a database and used to identify these objects in the future. One approach is to find all features in an image and compare them with those in the database. This comparison is done by using the Euclidian distance as metric and searching a k-d tree (with $d = 128$). In order to make this approach robust, each object needs to be identified by at least three independent features. For this, each descriptor stores the location, scale, and orientation of it relative to some common point on the object. This allows each detected feature to "vote" for the position of the object that it is most closely associated with in the database. This is done using a Hough transform.

For example, position (two dimensions) and orientation (one dimension) can be discretized into bins (30 degree width for orientation); bright spots in Hough-space then correspond to an object pose that has been identified by multiple features. Another popular approach uses the bag of words (BoW) technique, in which features are collected into groups that compose a "word." The words are then matched against query features to determine the similarity between the collected features and the query features and thus give a measure of likelihood that the object in the image is that of the query.

9.5   Feature Detection and Machine Learning

This chapter has introduced a variety of algorithms that turn high-dimensional input data into low-dimensional features, which can then be used to further reason about a problem. Recent advances in artificial neural networks (chapter 10) not only allow us to automatically train such feature detectors from data but also often outperform hand-coded feature detectors such as SIFT. Whereas the last decades have been dominated by hand-coding image understanding pipelines consisting of filtering, feature detection, and thresholding (see also chapter 8), modern neural network–based pipelines perform all these steps in the different layers of their network architecture. Like with low-level preprocessing (chapter 8), understanding basic feature detection algorithms remains important to understand what the different components of a neural network actually do, as well as how to deal with data for which no training information is available.

Take-Home Lessons

1. Features are "interesting" information in sensor data that are robust to variations in rotation and scale as well as noise.

2. Which features are most useful depends on the characteristics of the sensor generating the data, the structure of the environment, and the actual application.

3. There are many feature detectors available, some of which operate as simple filters, while others rely on machine learning techniques.

4. Lines are among the most important features in mobile robotics since they are easy to extract from many different sensors and provide strong clues for localization.

Exercises

1. Think about what information would make good features in different operating scenarios: a supermarket, a warehouse, a cave.

2. What other features could you detect using a Hough transform? Can you find parameterizations for a circle, a square, or a triangle?

3. Do an online search for SIFT. What other similar feature detectors can you find? Which provide source code that you can use online?

4. A line can be represented by the function $y = mx + c$. Then, the Hough-space is given by a 2D coordinate system spanned by m and c.

a) Think about a line representation in polar coordinates. What components does the Hough-space consist of in this case?

b) Derive a parameterization for a circle and describe the resulting Hough-space.

5. Implement a detector for the various targets in Ratslife. Start with basic 2D images, then think about what you need to change in order to find targets in any possible orientation.

6. Simulate, build, or get access to a range finder. Can you write an algorithm that reliably detects corners and openings?

10 Artificial Neural Networks

Artificial neural networks (ANNs) are part of a class of machine learning techniques that are loosely inspired by neural operation in the human brain; in robotics, they are generally used to classify or regress data for the dual purposes of perception (e.g., chapters 8 and 9) and control (e.g., as covered in chapter 11). While ANNs for the longest time have been just one of the many methods available to roboticists from the neighboring field of machine learning, recent advances in computing—in particular, graphical processing units (GPU)—and the availability of large datasets have enabled the training of neural networks with many layers, commonly referred to as *deep learning*. These (often massive) networks have led to revolutionary results in many fields including computer vision, natural language processing, video and speech processing, and robotics. Not too long ago, neural networks were considered "deep" if they had just more than two layers. Today, "deep" neural networks can have hundreds of layers and thousands of inputs and outputs—if not more. This is still shy of the human brain, which contains approximately 100^{11} neurons, each with thousands of synapses connecting a single neuron to thousands of others.

Remember: A *classification* problem requires that input data be classified between two or more classes; a *regression* problem requires a prediction of a (possibly continuous or high-dimensional) quantity. While a regression problem can be converted into a classification one (and vice versa), the machine learning community generally considers them two separate applications, and different techniques are developed to perform in each of these domains.

Machine learning is a large field that shares many of its foundations with robotics, in particular concerning probability theory and statistics. Deep learning may be used as a drop-in replacement for sensor preprocessing and conditioning, computer vision and feature extraction, and localization; it may even replace controllers for locomotion and grasping. For each of these applications, it is important to understand when deep learning may perform better than traditional approaches and when it does not. In a nutshell, deep learning models become first choice when not enough information exists to model a system using first principles. While a "deep enough" model with the right architecture might approximate any existing function in robotics, deep learning models lack "explainability" beyond statistical accuracy—that is, we may not easily be able to know how the approach actually works (in terms of which criteria it uses for its decisions) and when it might fail, usually making it a second choice behind an approach based on first principles with clear decision-making rationale.

The goals of this chapter are to introduce the following:

• Basic neural networks from the simple perceptron to multilayer neural networks

• Different network architectures and encodings to tackle a variety of regression and classification tasks

• Convolutional neural networks—including padding, striding, pooling, and flattening—and how they can be used to process spatial and temporal data

• Recurrent neural networks that introduce memory for classifying temporal data and perform control tasks

10.1   The Simple Perceptron

Artificial neural networks are inspired by neurons and synapses in the human brain and have been studied since the 1950s. One of the earliest models is the *perceptron*, which can classify an input vector x of dimension m into two classes. Such a problem is shown in figure 10.1. Variations of the simple perceptron remain the basic elements of deep neural networks until today. As detailed in figure 10.2, a perceptron has m inputs x_1 to x_m—each modulated by a weight w_1 to w_m as well as a threshold b—and it outputs either zero or one.

The perceptron classifies whether x lies above or below a hyperplane defined by the weights $w = \{w_1, \ldots, w_m\}$ using the following equation:

$$f(x) = \begin{cases} 1 & wx + b > 0 \\ 0 & otherwise \end{cases}. \tag{10.1}$$

Here, $wx = \sum_{i=1}^{m} w_i x_i$ is the dot product, and the nonlinear activation function $f(x)$ is also known as *Heaviside step function*. In practice, we are appending the value of "1" to the vector x so that $x_0 = 1$, which simplifies $wx + b$ (with $w = \{w_1, \ldots, w_m\}$) to wx with $w = \{w_0, w_1, \ldots, w_m\}$ where w_0 takes the role of b. This is illustrated in figure 10.2, where the bias b is alternatively labeled by w_0 and input $x_0 = 1$.

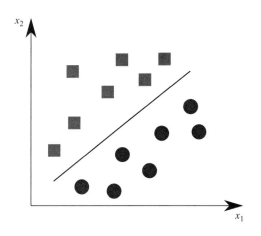

Figure 10.1
A two-dimensional dataset, where every element has two values (x_1 and x_2) and belongs to one of two classes (square and circle). In the simplest case of linear separation, it is possible to separate the two classes with just a straight line.

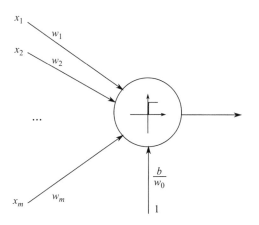

Figure 10.2
The simple perceptron passes the dot product between the inputs x and weights w through a Heaviside function, returning 1 when $wx + b > 0$ and 0 otherwise.

10.1.1 Geometric Interpretation of the Simple Perceptron

If w really defines a hyperplane, we should be able to easily visualize it when $m = 2$. When $m = 2$ (i.e., every data point x has only two dimensions), the separating hyperplane is a line such as the one shown in figure 10.1. Indeed, we can easily demonstrate this. Writing the dot product out yields

$$w_1 x_1 + w_2 x_2 + b = 0. \tag{10.2}$$

As we plot x_1 along the x-axis and x_2 along the y-axis, we can write

$$w_1 x + w_2 y + b = 0. \tag{10.3}$$

This can be rewritten into

$$y = -\frac{w_1}{w_2}x - \frac{b}{w_2} \tag{10.4}$$

and displayed within a scatter plot.

10.1.2 Training the Simple Perceptron

Training the perceptron equates to finding appropriate values for w and b that separate the data into two classes. This process can be performed iteratively:

1. Initialize all the weights with zeros or a small random number.
2. Compute the prediction $y_j = f(wx_j + b)$ for each data point x_j. A suitable choice for $f()$ is the Heaviside step function, as indicated in equation (10.1).
3. Calculate the mismatch between prediction y_j and the true class d_j to update the weights

$$w(t+1) = w(t) + r(d_j - y_j) * x_j. \tag{10.5}$$

4. Repeat steps 2 and 3 until a termination criteria (e.g., a decreasing error or maximum number of iterations) is reached.

Although it is very simple, this learning algorithm has still a lot in common with state-of-the-art algorithms. First, weights are updated in an iterative process using small increments governed by the parameter r, which is referred to as the *learning rate*. By changing w in small increments, the algorithm is literally rotating and translating the separating line in a direction that minimizes the *loss*, given by $d_i - y_i$. One can easily see that if the learning rate is set to too low of a value, the algorithm will never find a good solution. One can also see that if the learning rate is too large, the line might move too much, "skipping" the configuration that achieves optimal separation.

It is worth noting that this simple implementation is a de facto implementation of gradient descent, in this case with a loss function of the form $(d_i - y_i)^2$ that can be minimized by moving against the direction of its gradient, here $2(d_i - y_i)$. Other examples of gradient descent can be found in section 3.4.

Second, the learning algorithm requires multiple presentations of the dataset, as the error is computed for every point in the dataset. The more the amount of data, the longer the training! In this case the increase in time is linear—which is also generally true for more complex and modern learning algorithms.

Third, the error between the prediction and the true class is only calculated based on the given training data. Even if we were to train with unlimited amounts of data points, it would still be difficult to generalize for new data and to know whether these new measurements will be distributed in a way that is representative of the training data.

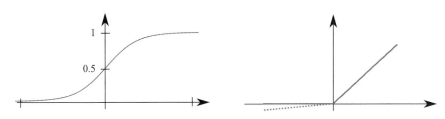

Figure 10.3
Typical activation functions used in neural networks: the sigmoid activation function (left), and the rectified linear unit or ReLU (right).

10.2   Activation Functions

Using an on-off Heaviside step function makes training a neural network using gradient descent rather difficult, as a function that switches from "not working at all" to "working completely" provides very little information concerning in which direction to move. It is therefore more desirable to have a smoother activation function. One such function is the *sigmoid function*:

$$\sigma(x) = \frac{1}{1 + e^{-x}}. \tag{10.6}$$

Its main characteristics are that it asymptotically stays between 0 and 1 and crosses the y-axis at 0.5. The sigmoid function is shown on the left side of figure 10.3.

The sigmoid function is attractive for learning because the direction in which the weights should move to improve the error is very clear in the vicinity of $wx = 0$, and computing its derivative is rather simple. While attractive in many cases, the sigmoid function has some drawbacks. For instance, when wx is very large or very small, the neuron either saturates or never activates—a phenomenon known as the *vanishing gradient* problem. Another drawback is that computing the sigmoid function is computationally expensive. An alternative is the hyperbolic tangent tanh(), which remains in the range of -1 to 1 and crosses the y-axis at 0.

A popular solution to decrease computation time is the *rectified linear unit (ReLU)*, which is given by

$$R(x) = max(0, x) \tag{10.7}$$

and is shown on the right side of figure 10.3. The dashed line indicates a refinement of the ReLU known as *leaky ReLU*, with a typical slope of 0.1; it improves learning for negative wx by providing a directional gradient.

Please note that we only talk about "perceptrons" when the Heaviside step function is used as an activation function.

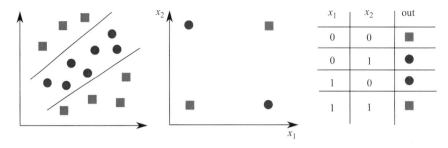

Figure 10.4
Data cannot be separated using a single line (left) in canonical form (center). This problem is known as the "XOR" problem because of the truth table of the associated classification problem (right).

10.3   From the Simple Perceptron to Multilayer Neural Networks

We have seen that the single perceptron is able to linearly separate a dataset, returning "0" or "1" as a function of the data being below or above the separating hyperplane defined by the weight vector w. However, it is easy to see that some problems cannot be linearly separated. In the example shown in figure 10.4, the square and the circle data points are not separable by a single line but require at least two lines. This problem is known as the "XOR" problem, which can be seen by looking at just four data points at $(0,0)$, $(0,1)$, $(1,0)$, and $(1,1)$. Tabulating this data together with its shape reveals a truth table with the characteristics of logical "exclusive or" (XOR)—for example, x_1 and x_2 have to be different for the output to be true (here "circle"), whereas the output is false (here "square") when the inputs are the same.

We already know that a single perceptron can create a single separating hyperplane; we will therefore need at least two perceptrons to solve the XOR problem. Using two perceptrons in parallel will yield tuples of the kind $(0,0)$, $(0,1)$, and so on; hence, we then need another perceptron to recombine these tuples into a single output. Figure 10.5 shows the simplest multilayer perceptron that can be trained for the XOR problem, with one *input layer*, a so-called *hidden layer*, and an *output layer*.

10.3.1   Formal Description of Artificial Neural Networks

As with the simple perceptron, we will use node i's bias as the $0-$th weight vector, that is:

$$w_{0,j}^k = b_j^k. \tag{10.8}$$

Here, we use the following notation: We denote the layer with a superscript and the index of the incoming node and the outgoing node with a subscript tuple. That is, $w_{i,j}^k$ is connecting the $i-$th incoming weight to the $j-$th node of the $k-$th layer (the $i-$th incoming weight is the $j-$th node in layer $k-1$). This, as well as the simple example network from above, is illustrated in figure 10.6. Each layer, denoted by the index k, has exactly r^k nodes.

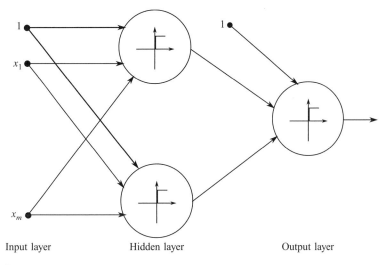

Figure 10.5
A simple multilayer perceptron with one input layer, one hidden layer, and one output layer.

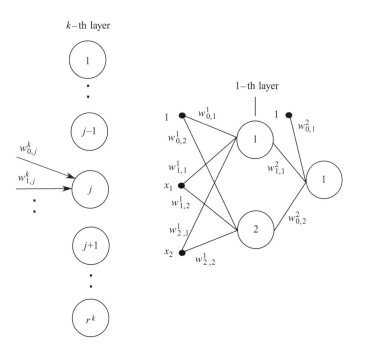

Figure 10.6
Notation used to index weights (left) with respect to layer k and the multilayer network from figure 10.5 (right).

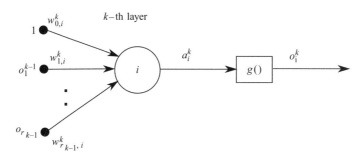

Figure 10.7
Inputs and outputs of neuron i in the k–th layer showing activation a_i and output o_i.

Inputs and outputs

The output o_i of node i is given by

$$o_i = g(a_i^k),$$ (10.9)

where $g()$ is a nonlinear activation function such as (but not limited to) the ones described in section 10.2 or the Heaviside step function. Here, a_i^k is known as the *activation*—that is, the weighted sum computed by node i in layer k:

$$a_i^k = \sum_{j=0}^{r_{k-1}} w_{j,i}^k o_j^{k-1},$$ (10.10)

with o_j^{k-1} the j–th output of the previous layer. This is illustrated in figure 10.7.

In case of k being the output layer, o_i^k should be equivalent to y_i^k. Likewise, in case of $k-1$ being the input layer, $o_i^{k-1} = x_i$.

10.3.2   Training a Multilayer Neural Network

Finding a set of weights and bias values, which is a few parameters for a simple two-dimensional problem but potentially billions for a "deep network," is an NP-complete problem (Blum and Rivest 1992). We therefore need an efficient approximation. To this end, we consider a training dataset consisting of input-output pairs x_i and y_i with $i = 1..N$, and a feed-forward neural network with parameters w.

Loss function

The goal of training is to minimize an error function such as the mean-squared error

$$E(x, y, w) = \frac{1}{2N} \sum_{i=1}^{N} (\hat{y}_i - y_i)^2$$ (10.11)

between the output $\hat{y}_i$ that the neural network with parameters w computes and the known value y_i from the *training set*. Similar to the perceptron, we can reduce $E(x, y, w)$ by iteratively descending along its gradient, using the following equation:

$$w(t+1) = w(t) - \alpha \frac{\partial E(x, y, w(t))}{\partial w}. \qquad (10.12)$$

This process is nontrivial, as calculating the partial derivatives across the computation graph of the neural network requires the chain rule. An algorithm known as *backpropagation* is described in appendix D.

10.4   From Single Outputs to Higher Dimensional Data

Extending a neural network from one single output to multiple binary classifiers is straightforward, requiring only an increase in the dimensionality of the output vector. Much less straightforward is encoding more complex data, which leads to the following question: How can we represent numerical values, such as digits from 0 to 9 or characters from A to Z?

One-hot encoding

A very common approach is known as *one-hot encoding (OHE)*. In OHE, n discrete labels such as numbers or characters are encoded as a binary vector of length n. To encode the $i-$th element of a set of labels, this vector is 0 except at position i. For example, to encode the characters $0 \dots 9$, OHE would represent them as follows:

$$0 = (1, 0, 0, 0, 0, 0, 0, 0, 0, 0)$$

$$1 = (0, 1, 0, 0, 0, 0, 0, 0, 0, 0)$$

$$2 = (0, 0, 1, 0, 0, 0, 0, 0, 0, 0)$$

$$3 = (0, 0, 0, 1, 0, 0, 0, 0, 0, 0)$$

$$4 = (0, 0, 0, 0, 1, 0, 0, 0, 0, 0)$$

$$5 = (0, 0, 0, 0, 0, 1, 0, 0, 0, 0)$$

$$6 = (0, 0, 0, 0, 0, 0, 1, 0, 0, 0)$$

$$7 = (0, 0, 0, 0, 0, 0, 0, 1, 0, 0)$$

$$8 = (0, 0, 0, 0, 0, 0, 0, 0, 1, 0)$$

$$9 = (0, 0, 0, 0, 0, 0, 0, 0, 0, 1).$$

Softmax output

Whereas OHE transforms the training input into a discrete probability distribution, nothing in the neural network will ensure that the data will also come out like that. A sigmoidal activation function would ensure that each value remains between 0 and 1, but a ReLU does not. We therefore need a final layer that ensures each output is limited to the range 0 to 1 *and* that the sum of all elements will add up to 1. This is usually achieved using a so-called *softmax* layer. The softmax function is given by

$$\sigma(\mathbf{z})_j = \frac{e^{z_j}}{\sum_{k=1}^{K} e^{z_k}} \quad for \quad j = 1, \ldots, K. \tag{10.13}$$

That is, a vector $z \in \mathbb{R}^K$ will be converted into a $K-$dimensional vector whose $j-$th element is given by the above formula.

So, why not just normalize with the actual values—that is, using z_j instead of e^{z_j} or, even easier, using $\arg\max_j$ to set the highest value of z to 1 and leave the rest at 0? The reason is that each layer needs to remain differentiable for backpropagation to work. Yet the "brutal" cutoff introduced by the arg max function is exactly what we want for the network to optimally match the training input. This is why the exponential function is used. It literally and exponentially emphasizes larger values over smaller values, making the class with the highest probability stand out.

10.5   Objective Functions and Optimization

The key idea to train neural networks is to change the network's parameters so that a certain objective function (called loss function) is minimized. This is usually done by evaluating the gradient of the objective function with respect to the network's parameters. Being differentiable is therefore a key requirement for a useful objective function. However, the magnitude of the weights can dramatically affect neural network performance, and finding this magnitude is entirely dependent on the type of learning problem.

10.5.1   Loss Functions for Regression Tasks

So far, we have considered the so-called *mean-squared error* (MSE),

$$E = \frac{1}{2N} \sum_{i=1}^{N} (\hat{y}_i(w) - y_i)^2, \tag{10.14}$$

which is the average error over a set of N pairs of predictions $\hat{y}$ that are dependent on the network parameters w and known values y (see also section 15.2.1). This function is particularly convenient because the square makes it convex, allowing to find its minimum by following its gradient ("gradient descent").

MSE is most suited for *regression* tasks in which data points are fitted to a model such as a line. Using a sigmoid or other continuous activation function, the error for each class can also be interpreted as a distance from the separating hyperplane, which makes MSE also suitable (but not optimal) for these kind of tasks. An example is illustrated figure 10.8.

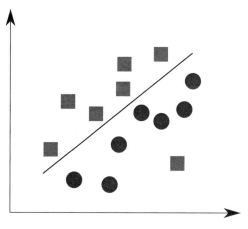

Figure 10.8
A regression problem with an outlier.

From figure 10.8 it is clear that MSE poorly deals with outliers. If one value deviates largely from the prediction, the quadratic term in MSE will heavily "punish" this value. An alternative to MSE is the *mean absolute error* (MAE):

$$E = \frac{1}{2N} \sum_{i=1}^{N} \|\hat{y}_i(w) - y_i\|. \tag{10.15}$$

Here, the absolute value ensures that the error is always positive no matter the direction, but large errors are weighted on the same order of magnitude as smaller ones. MAE is therefore better suited if your training set contains outliers.

In practice, a large variety of loss functions have been developed to combine features of both MSE and MAE; in the simplest form of the *Huber loss* function, this is achieved via a simple piecewise combination.

10.5.2 Loss Functions for Classification Tasks

Although a classification task can be cast into a regression problem, classifying is more akin to throwing a dice. Indeed, the output of the softmax layer is a discrete probability distribution in which each element $y_i = (p_0, \ldots, p_c, \ldots, p_N)$ is the probability of an instance x_i to be of class c in N classes total.

We speak of the *entropy* of a probability distribution as the amount of "variety" that we expect. To make an example, a uniform distribution has the highest entropy because of the high number of possible outcomes, whereas the one-hot encoded vectors are probability distributions with very low entropy. The entropy of the distribution of y_i (the training vector that stores the true class c for each instance x_i) is given by

$$H(y_i) = -\sum_{c=1} Np_c \log p_c. \tag{10.16}$$

Here, the logarithm can be of basis ten or two. In any case, the entropy function has a couple of interesting properties: First, the logarithm from 0 (negative infinity) to 1 is negative (this is why probabilities yield positive values). Second, the logarithm of 1 is zero—that is, a distribution with only one element ($p_c = 1$) has the lowest possible entropy. Third, the lower the individual entries for p_c are—for example, in a uniform distribution where $p_c = \frac{1}{N}$—the higher the entropy.

In every dataset, there will always exist a true distribution $P(C = i)$ of your data. By classifying every element in the training set, the neural network also generates its own distribution or "interpretation" of the data. Ideally, in the case of a 100 percent fit, the neural network will generate (or "learn") the exact same distribution as the one that describes the training set. In the worst case, the network will generate a distribution that is completely different. Evaluating a neural network's performance is therefore a matter of comparing two probability distributions.

One way to compare two distributions is to use their entropy—a process known as *cross-entropy*, which is written as

$$H(\hat{y}, y) = -\sum_{i=1} Ny_i \log \hat{y}_i, \tag{10.17}$$

with $y_i = p_i$ being the known probability for instance x to be class i and $\hat{y}_i$ being the prediction. As the neural network will never perfectly represent the data, the cross-entropy will always be larger than the entropy of the true distribution. That is:

$$H(y) - H(\hat{y}, y) \leq 0. \tag{10.18}$$

This difference between the entropy of the true distribution and the cross-entropy between the true and the estimated distribution is known as *Kullback-Leibler divergence*. It is a measure of dissimilarity between two distributions.

10.5.3   Binary and Categorical Cross-Entropy

In the case where there are only two classes, the *binary cross-entropy* is calculated as follows:

$$H(\hat{y}, y) = -\sum_{i=1}^{N} y_i \log(\hat{y}_i) = -y_1 \log(\hat{y}_1) - (1 - y_1) \log(1 - \hat{y}_1). \tag{10.19}$$

As there are only two classes (either true or false), $\hat{y}_2$ directly follows from $1 - \hat{y}_1$. The more general case for $N > 2$ is known as *categorical cross-entropy*. When using one-hot encoding, only class c has probability 1 ($y_c = 1$), reducing the cross-entropy to

$$H(\hat{y}, y) = -\log(\hat{y}_c). \tag{10.20}$$

with c the true class (the other terms are zero). Combined with the softmax activation function, the categorical cross-entropy therefore computes as

$$H(\hat{y}, y) = - \log \left(\frac{e^{\hat{y}_c}}{\sum_j^N e^{\hat{y}_j}} \right). \tag{10.21}$$

10.6 Convolutional Neural Networks

A drawback of the ANN architectures that we have covered so far is that they do not consider the spatial information that might be hidden in a dataset. For example, as detailed in chapter 8, in the context of vision, it is important to interpret the value of a certain pixel depending on what can be seen nearby: A blue pixel surrounded by white ones might be an eye, whereas a blue pixel surround by blue ones might be an ocean. In addition to color, neighboring pixels also encode structure. When looking at the MNIST dataset (a collection of hand-drawn numbers from zero to nine), we might, for example, be looking for crosses (such as the center of an eight), T-shaped junctions (such as in the letter four) or half-circles (like in the letter three), whose number might then serve as features for our neural network. The scale-invariant features (SIFT) in chapter 9 were a good example of a hand-coded approach to encode such spatial information. We will now see how ANNs can find such features automatically.

If you recall, one way to extract features in image processing is by *convolving* an image with a *kernel* (see, for example, a convolution with a 3×3 and a 7×7 kernel in figure 10.9). During a convolution, the kernel is swept across the input image, summing a piecewise multiplication of each element of the kernel with the underlying image pixels (see also chapter 8). As all multiplications are summed, such an operation yields only one pixel. As the kernel has to start somewhat inside the image (unless its borders are padded with appropriate values), we are losing half the width of the kernel on each side. In the example above, a 3×3 kernel turns a 28×28 input image into a 26×26 output image, and a 7×7 kernel turns it into a 22×22 pixel image. Mathematically, the convolution is defined as

$$x(n_1, n_2) * h(n_1, n_2) = \sum_{k_1=-\infty}^{\infty} \sum_{k_2=-\infty}^{\infty} h(k_1, k_2) x(n_1 - k_1, n_2 - k_2), \tag{10.22}$$

where bounds (here, infinity) need to be chosen so that the kernel starts at the upper left corner of the image and ends at the lower right corner. It is also possible to artificially grow the input image by adding pixels around it, which is known as *padding*. Note that the resulting output is identical to examples shown in chapter 8.

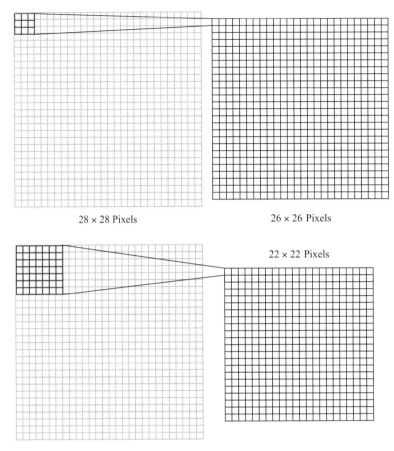

28 × 28 Pixels   26 × 26 Pixels

22 × 22 Pixels

Figure 10.9
Convolution with a 3 × 3 and a 7 × 7 kernel and resulting reduction in image size.

10.6.1   From Convolutions to 2D Neural Networks

When looking at how one individual pixel in the output above gets computed, we assume that the input pixel is labeled $x_{i,j}$ with i the row and j the column of this pixel. We also assume the entries of the convolution kernel to be indexed in a similar way. Using a 3×3 kernel, the first pixel of the output is then calculated by

$$o_{0,0} = \ x_{0,0}w_{0,0} + x_{0,1}w_{0,1} + x_{0,2}w_{0,2} \qquad (10.23)$$
$$+ x_{1,0}w_{1,0} + x_{1,1}w_{1,1} + x_{1,2}w_{1,2}$$
$$+ x_{2,0}w_{2,0} + x_{2,1}w_{2,1} + x_{2,2}w_{2,2}.$$

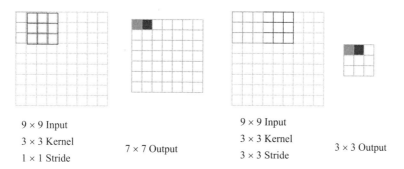

9 × 9 Input
3 × 3 Kernel   7 × 7 Output
1 × 1 Stride

9 × 9 Input
3 × 3 Kernel   3 × 3 Output
3 × 3 Stride

Figure 10.10
Convolution with 1 × 1 and 3 × 3 stride and resulting output.

This operation is therefore simply computing the dot product of the value of nine pixels with the kernel weights. Adding a bias value and an activation function such as ReLu is therefore identical to adding a hidden layer with nine neurons!

Performing the convolution by moving the convolution kernel with a width of $(2r + 1)$ across an entire $X \times Y$ image is therefore akin to creating $(X - 2r)(Y - 2r)$ "convolutional" neurons; the resulting structure is called a *feature map*. Note that the "weights" of the feature map—that is, the entries of the kernel matrix—are identical for each neuron in the feature map. We can now repeat this step with additional kernels, resulting in multiple feature maps, which then form a *convolutional layer*.

Importantly, as this structure is very similar to the conventional neural network structure (except for the fact that a large number of weights are identical), the parameters of each kernel can also be trained using backpropagation (see appendix D).

10.6.2 Padding and Striding

As mentioned earlier, a convolution of kernel width $2r + 1$ reduces the input by r on each side. If this is not desired (e.g., when multiple convolutional layers are used in series), *padding* can be used to surround the input image with up to r pixels, which results in the output image having the same dimension as the input image. Instead of moving the convolution kernel pixel by pixel, skipping pixels will further reduce the size of the output image. The amount by which the convolution kernel is moved is known as *stride*. This is illustrated in figure 10.10 for strides of one and three.

10.6.3 Pooling

The feature maps that result from convolution each identify specific features that are defined by their kernels. Through training, it is possible to identify these kernels and specialize them for specific characteristics that are of interest. Some might "fire" on edges, others on intersections of lines, and others on very specific patterns in the dataset. Activation

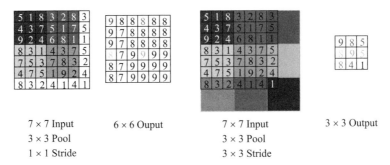

7 × 7 Input	6 × 6 Ouput	7 × 7 Input	3 × 3 Output
3 × 3 Pool		3 × 3 Pool	
1 × 1 Stride		3 × 3 Stride	

Figure 10.11
Pooling using a pool size of 3 × 3 for different strides and corresponding output.

functions may be used to further amplify this effect, making a clear distinction between whether a feature is present or not. However, in most practical applications such features are rather sparse, and whether they exist in a larger area or not might be the most salient information. This can be achieved by a *pooling layer*.

A pooling operation applies a window to select the maximum (in which case it is referred to as *MaxPooling*) or the average, among many other possible nonlinear functions, from a window of a given size. Figure 10.11 shows the result of a MaxPooling layer with pool size of 3 × 3 and stride lengths of 1 × 1 and 3 × 3. Usually, the stride length is the same as the width of the window.

Although the *max*() function is not differentiable, MaxPooling can still be used in back-propagation by selectively passing the gradient to only the neuron that has shown to have the maximum activation and setting the gradient of all other neurons to zero. When an averaging pooling function is used, the gradient is divided among all neurons in the pool in equal parts.

10.6.4   Flattening

The first step in previous neural network models has been to flatten a two-dimensional (2D) input image into a one-dimensional (1D) vector. This has been the precondition to apply a dense layer and has been accomplished during preprocessing. However, convolutional neural networks (CNNs) require multidimensional inputs (e.g., 2D images with multiple color channels). Turning a multidimensional tensor into a vector is known as *flattening* and results in simple reordering. For example, a red, green, blue (RGB) image of dimensionality (28 × 28 × 3) might be turned into 20 convolutional filters, or 2,352 individual neurons. A flattening layer arranges them again in a single vector.

10.6.5   A Sample CNN

Figure 10.12 shows a typical CNN that combines multiple convolutional and pooling layers. The network takes a 28 × 28 image as an input and trains 20 different 5 × 5 convolution

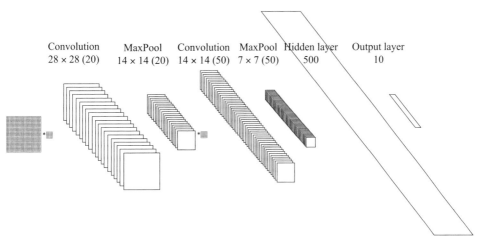

Convolution
28×28 (20)

MaxPool
14×14 (20)

Convolution
14×14 (50)

MaxPool
7×7 (50)

Hidden layer
500

Output layer
10

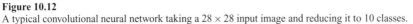

Figure 10.12
A typical convolutional neural network taking a 28×28 input image and reducing it to 10 classes.

kernels to create 20 feature maps of 28×28 each. This layer is followed by a MaxPooling layer that downsamples each feature map by a factor of two. These feature maps are then convolved with 50 5×5 convolution kernels to create 50 14×14 feature maps. These will again be downsampled by a MaxPooling operation. The resulting 50 feature maps are then flattened and fed into a hidden layer of 500 neurons and finally into a softmax-activated output layer with 10 neurons.

10.6.6 Convolutional Networks beyond 2D Image Data

Convolution kernels emphasize areas of similarity. This can be readily understood when considering a simple kernel like $[[0, 9, 0], [0, 9, 0], [0, 9, 0]]$, which emphasizes vertical lines but ignores horizontal ones. Training a convolutional network therefore automatically finds regularities in the training set as well as in the resulting feature map—often generating hierarchical representations by itself. A common example is a convolutional neural network for face detection in which early layers detect low-level features, which then get recombined into noses, ears, mouths, and eyes in deeper layers.

Convolutional neural networks are not limited to 2D image data. They can also be applied to 1D time series. Here, the CNNs will find distinct patterns—for example, peaks in an accelerometer or gyroscope reading, which can then be used collectively to classify complex signals.

10.7   Recurrent Neural Networks

So far, we have only worked with static data. Even if data had a temporal nature, we have simply concatenated inputs and looked at a piece of history all at once. When using a dense

network, all inputs are initially of equal importance, and it is up to the network to identify salient information. Although convolutional layers might help to dictate some sense
of order—a 1D convolutional layer might as well be interpreted as detecting a pattern in
a time series—dense layers focus on the values of individual features, not on the order of
information.

For example, it is straightforward to train a neural network controller to transform input
data from sensors into motor commands to perform tasks such as light following, obstacle
avoidance, and wall following, as described in chapter 11; however, such a controller will
be purely reactive and not be able to, for example, escape a U-shaped obstacle.

To overcome this limitation, it is useful to introduce a notion of state in a neural network.
In this case, the detection of an event such as "getting stuck" may be used to modify the
network state in some way. This is accomplished using so-called *recurrent neural networks*.
A recurrent neural network uses a special kind of neuron that sums the input x_t at time t
with the value of the hidden state h_{t-1} at the previous time-step $t-1$ to compute a hidden
state h_t at time t. Both terms of this sum are weighed by weights W and U. The output
of the recurrent layer is the hidden state h_t weighed by a third weight V and run through
a second activation function. The equations below shows the computation of a recurrent
neural network (RNN) layer in vector form, passing the hidden states through a softmax
activation:

$$h_t = \tanh(Wh_{t-1} + Ux_t) \tag{10.24}$$

$$y_t = softmax(Vh_t). \tag{10.25}$$

This relationship is shown in figure 10.13. As an RNN cell is reusing its internal state
h_t in the next iteration, a network that looks back N time-steps is modeled as N cells that
are laterally connected. As this is how an RNN is actually implemented, the data from all
time-steps are presented at the same time.

Take-Home Lessons

• Artificial neural networks and the tools associated with them have become a powerful tool
to skip modeling a system using first principles and to simply learn its properties from data.
As such, they are capable of replacing many of the models discussed in previous chapters,
ranging from kinematics to vision, feature detection, and controls.

• Simple neural networks are capable of both classification and regression akin to techniques described in chapter 9, whereas convolutional networks are capable of filtering and
preprocessing techniques such as described in chapter 8.

• When a system is not purely reactive but requires state (described in chapter 11), recurrent
neural networks are needed to implement a notion of memory.

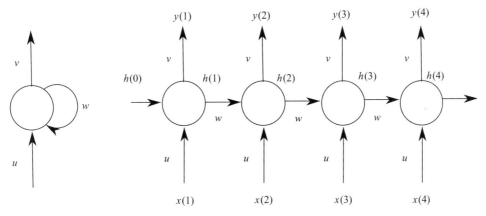

Figure 10.13
A sample recurrent neural network (left) and its expanded version (right) that is looking back four time-steps.

Exercises

1. Implement the simple perceptron training algorithm and use it to find a separating hyperplane for simple data.

2. Find out how to implement the auto differentiation (or auto gradient) function in your favorite numerical package (e.g., *NumPy* or *PyTorch*) to automatically calculate the derivative of your loss function.

3. Use a machine learning package of your choice to train a classifier for synthetic images such as the "Ratslife" landmarks. If you can, use a real robot to generate appropriate training data.

4. Select a simple 2D target (e.g., a cross on white background) and record images from different distances and angles. Can you train a CNN to predict these two quantities from your image?

5. Select a pretrained image classifier from your preferred machine learning toolkit and use it as the basis to train your classifier for either landmark recognition or pose recognition. How does using a pretrained classifier affect learning time and accuracy?

6. What kind of network architecture would you chose to track the robot's location (odometry) based on encoder inputs?

7. Download the "Robot Execution Failures Data Set" from the UCI Machine Learning Repository (Dua and Graf 2019). It contains time-series data from a robot's force-torque sensor and whether manipulation was successful. Define a recurrent neural network architecture for this data and train it.

11 Task Execution

In their most basic implementation, sensors and actuators can be directly tied to each other, removing the need for computation. Such robots are purely reactive, thereby missing the ability to "think" or plan. In order to achieve more complex behavior, memory and state are needed to switch between different controllers and algorithms.

This chapter introduces these basic principles as well as their implementation, starting with basic reactive controllers (section 11.1), then moves to more advanced concepts that let the robot make basic "if" ... "then" decisions using finite state machines (sections 11.2 and 11.3) and finally concepts such as behavior trees and semantic planning (sections 11.4 and 11.5). This chapter covers the following topics:

- The basic control loop that allows robots to react to their environment
- Ways to introduce state that allow robots to switch their behavior
- Basic concepts that allow the robot to reason about its discrete states and pick the next action

11.1 Reactive Control

A wide variety of robotic behaviors can be accomplished by directly connecting sensor input to actuator output. These behaviors can even be accomplished without a computer by using analog electronics that provide appropriate conditioning. Simple autonomous robots that use these concepts have been demonstrated as early as the 1950s (Walter 1953) and have become known as "tortoises." For example, by tying the output of a light sensor to a motor controller, the motor turns faster when the light is brighter. Using an inverse relationship between sensor and motor, the motor will turn slower when the light is brighter. When used in a differential-wheel configuration with two motors and two light sensors, such as in figure 11.1, such a robot can be built to either drive toward or away from the light.

Formally, we can express the light-following behavior—also known as *phototaxis*—using the relationship between the left and right wheel speeds $\dot{\phi}_l$ and $\dot{\phi}_r$ and the measurements of

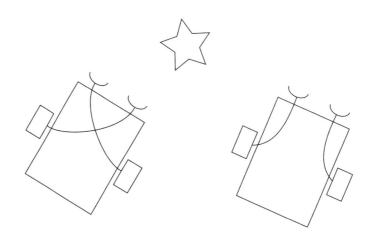

Figure 11.1
Two vehicles approaching a light source. The brighter the light, the more each motor turns. The left vehicle will therefore approach the light by turning toward it, and the right vehicle will avoid it by turning away from it.

the right and left light sensors λ_r and λ_l:

$$\dot{\phi}_l = a\lambda_r + b \tag{11.1}$$

$$\dot{\phi}_r = a\lambda_l + b, \tag{11.2}$$

with a being a constant weight and b a bias term. We observe that the left wheel turns faster the brighter the light shines on the right sensor. If the right light sensor receives more light than the left sensor, the right wheel will turn slower, thereby exhibiting a phototaxis behavior that results in a right turn.

A more complex reactive behavior is obstacle avoidance. Assuming the output of an obstacle sensor increases as the obstacle nears (e.g., an infrared proximity sensor), we can use the same principle to compute the wheel speeds such that the obstacle is actively avoided. An example for a differential-wheel robot with eight infrared proximity sensors is illustrated in figure 11.2 and given by:

$$\dot{\phi}_l = -6d_0 - 6d_1 - 19d_2 - 13d_3 + 94d_4 + 63d_5 - 50d_6 - 6d_7 + b$$

$$\dot{\phi}_r = -6d_0 + 50d_1 + 63d_2 + 94d_3 - 22d_4 - 10d_5 - 6d_6 - 6d_7 + b,$$

with the eight sensors $d_0 \ldots d_7$ arranged similarly to the e-Puck differential-wheel robot in figure 11.2, with d_0 as the left rearward sensor and the other sensors arranged clockwise such that d_7 is the right rearward one.

Behaviors such as phototaxis and obstacle avoidance can also be combined by simply weighing each input accordingly. This idea has been popularized by the neuroscientist Valentino Braitenberg, who augmented this system with additional ideas around basic forms

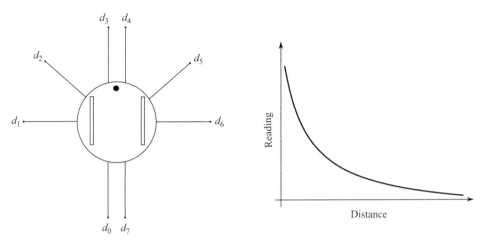

Figure 11.2
A schematic of an e-Puck, a differential-wheel robot with eight infrared distance sensors (left) and typical sensor response as a function of distance (right).

of learning (by changing the weights based on events such as collisions), natural selection (building robots with random weights and selecting those that perform best), and analogies to the human brain (Braitenberg 1986). Controllers of these kind are therefore often called "Braitenberg vehicles." Indeed, the controllers bear strong resemblance to artificial neural networks such as those described in chapter 10, and "optimal" values to obtain a certain behavior can be obtained using evolutionary computation (Floreano and Mondada 1998) or by training a neural network on recordings of input/output pairs that correspond to a desired behavior.

There are numerous variants of the control architecture including the *subsumption archi-tecture* (Brooks 1990) and *motor schemas* (Arkin 1989) that propose variations of switching different components of a reactive controller on and off to obtain desired behaviors. However, while useful for achieving relatively simple behaviors and capable of exhibiting more complex, emergent ones, these approaches are difficult to manage in practice and are better solved by being embedded in high-level control frameworks.

11.1.1   Limitations of Reactive Control

The limitations of a reactive control scheme are evident when considering that a robot combining both phototaxis and obstacle avoidance will still get stuck if presented with a U-shaped obstacle (figure 11.3). While obstacle avoidance will prevent the robot from hitting the obstacle, as soon as the way is clear, the robot will keep turning toward the light, thereby getting stuck in a loop. This type of behavior can also be observed in insects such as flies or moths.

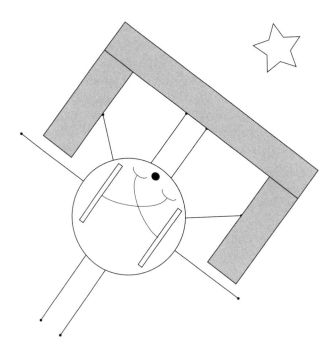

Figure 11.3
A differential-wheel robot with distance and light sensors wired in a "light following" configuration in a U-shaped obstacle. Although the obstacle will be avoided, the behavior to move toward light will continuously drive the robot into the obstacle unless state is added.

In order to avoid this situation, the robot needs to memorize its previous state and switch behaviors accordingly. For example, in addition to the basic combined avoidance and following behavior ("avoid and follow"), we can introduce an additional term ("wall following') in which the robot uses its proximity sensors to maintain a constant distance from a wall. To switch from one to the other behavior, we need to change the constant gains into dynamic ones that change their value based on other observations the robot makes. For example, the robot could estimate its progress by monitoring whether its light sensor is constantly increasing, and if it is not, inhibiting phototaxis behavior and emphasizing wall following.

Designing reactive systems with time-dependent behavior and state is potentially realizable with very simple electronics; for this reason, we have seen real-world implementations of such mechanisms in the form of robot vacuum cleaners. However, it very quickly becomes difficult to manage. It is therefore desirable to establish discrete abstractions for various behaviors, which can be more easily managed and understood by a programmer.

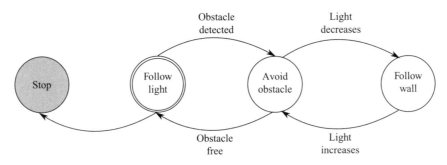

Figure 11.4
A simple finite state machine (FSM) with four states. The final state is colored in gray; the initial state is set apart with a double circle.

11.2   Finite State Machines

A simple yet powerful tool to facilitate switching between different behaviors is a so-called *finite state machine* (FSM). In an FSM, each state is associated with a specific controller. In practice, an FSM consists of a global variable that stores the current state and a series of "if" statements that contain the code that is associated with each unique state. For example, an FSM to perform phototaxis while avoiding U-obstacles could consist of four states, one for each desired behavior: one state that computes wheel speeds so that the robot moves toward the light, another to use its sensors to avoid obstacles ahead of it, a state that computes wheel speeds so the robot performs wall following for a fixed amount of time, and finally a state to stop the robot. An example of these states is shown in figure 11.4.

To specify an FSM, one also needs to specify the *state transitions*—that is, the conditions that determine when to switch states. For example, if multiple sensors detect an obstacle (implying that it may be a large one), then it may be desirable to have the FSM transition from its first state (phototaxis with simple obstacle avoidance) to its second (avoid obstacles). Should the light measurement decrease, such as when the robot needs to make a U-turn to avoid the obstacle, the behavior should transition to wall following. Once the light increases again, such as when the robot has gone around the obstacle, it resumes light following. Once the light sensor exceeds a threshold ("bright enough"), the robot stops.

Finally, it is necessary to specify an initial state (the state the system starts in) and any number of final states (terminal states that signify program termination). In the example in figure 11.4, the program will always start in light-following mode and terminate once the state labeled "stop" is reached.

Formally, a FSM is defined by a tuple $(\Sigma, S, s_0, \delta, F)$ where:

- Σ is the input *Alphabet* (i.e., a set of symbols that represent events that can trigger state transitions).

- S is a finite set of states.
- s_0 is an initial state and an element of S (i.e., $s_0 \in S$).
- δ is the state-transition function $\delta : S \times \Sigma \to S$ that maps combinations of states in S and symbols x in Σ to a new state in S.
- F is the set of final states, a subset of S.

Historically, this definition stems from FSMs formally defining the working of a computer with a stream of symbolic commands of an actual program. In robotics, symbols that trigger state transitions can themselves be the result of complex computations. An example is the robot switching to wall-following behavior if it has not made actual progress toward its goal in some time and resuming phototaxis once it reaches a position that is closer to the light than it was before.

In conjunction with a controller for each state, an FSM is called a *hybrid system* (Van Der Schaft and Schumacher 2000) as it combines both discrete (the state) and continuous (the controller outputs) variables.

11.2.1   Implementation

A low-level robot controller is usually implemented as a loop with fixed loop time, for example 100 milliseconds (*ms*) for slow-moving differential-wheel robots or 1 ms for dynamical systems such as drones or humanoid robots. At each start of the loop, the controller reads all sensors, then branches into the part of the code that corresponds to its current state, processes sensor information, computes actuator output, and finally sends the control commands to the actuators.

Unlike a computer program that can process information as fast as possible, a robot controller needs to wait until sensor information is actually available and actuator commands are executed (i.e., the robot has physically moved). As the robot keeps moving while computation is ongoing, it is important to run the main loop at a constant rate. As computation is usually much faster than the loop time, it might be necessary to use an internal clock to wait until the loop time is completed.

It is helpful to capture all the FSM's state transitions in a drawing as shown in figure 11.4. In practice, FSMs are difficult to develop, debug, and maintain. As the controller is being developed and experimented with, edge cases require the addition of transitions and states. With N states, there are possibly $N \times N$ state transitions, and it is typical to discover necessary additional state transitions as the FSM is specified. FSMs with many transitions become difficult to depict graphically, making it difficult to visualize what the program will actually do.

Whenever a state is added or removed, the programmer has to identify transitions required for the new one or modify all other states that have transitions to the one being removed, further contributing to FSM maintenance difficulty. Although behaviors such as obstacle avoidance are generic, each state also contains transitions that are specific to an application, potentially making it difficult to reuse states in other FSMs (modularity). FSMs also have difficulties with state transition conditions that cannot be evaluated in a single time-step (e.g., when averaging the gradient of the light sensor to robustly detect an increase or decrease). In this case, these computations need to be carefully woven into the state execution code, adding complexity and making maintenance difficult.

11.3   Hierarchical Finite State Machines

To make FSMs more manageable and to deal with information that needs to be pro-cessed at different time-scales, states can be grouped into clusters—each with their own associated FSM, thereby creating "super-states" organized in hierarchical fashion. This construct is usually referred to as hierarchical FSM (HFSM) but also known as "statechart" (Harel 1987). Considering the example in figure 11.4, each state might as well be a super-state (e.g., the "follow wall" state may consist of an FSM that deals with an edge case such as rounding sharp corners). An example HFSM is depicted in figure 11.5. State transitions between super-states can be tied to states in the included FSM or implicitly connected to all states of the included FSM, which allows leaving the super-state from every state therein.

Super-states can also be executed in parallel, providing events that lead to state transitions in other FSMs. For example, detecting whether a robot still makes progress toward a light while avoiding an obstacle might require computing a running average and dropping out-liers. This is illustrated by two super-states, one for the actual light-following behavior and the other for computing a running average of the light measurement, rejecting outliers, and generating symbols that can be consumed by the light-following behavior. In figure 11.5, we are using the character "/" to delineate state transition conditions such as "Light decreases" and the symbol that is generated during the state transition (here "LD"). These symbols can then drive state transitions in other clusters.

11.3.1   Implementation

In practice, HFSMs are implemented in distinct processes that run independently and asyn-chronously. They can communicate using an inter-process communication (IPC) framework such as XMLRPC or REST, which are socket-based networking protocols that allow one to exchange eXtended Markup Language (XML) or JavaScript Object Notation (JSON) data structures between two processes on the same or different computers using a network-ing interface. There exist many IPC frameworks that are particularly targeted at robotics and introduce abstractions for robot-specific data structures such as coordinate frames or video streams, the associated tools to manage them, and bindings for different languages to publish and subscribe to them. A prominent example is the Robot Operating System (ROS).

HFSMs solve some of the problems of FSMs by increasing modularity and simplify-ing programmability, but they still have the problem that N states can lead to N^2 state transitions, each of which needs to be manually coded.

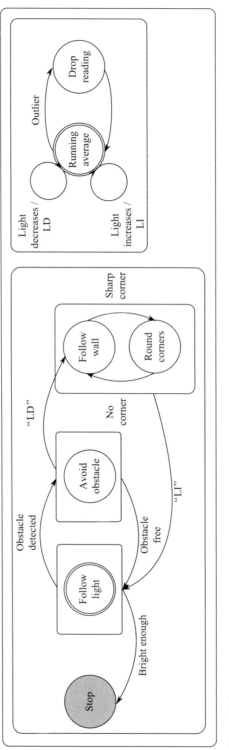

Figure 11.5
A hierarchical FSM with the states from figure 11.4 as super-states, a more sophisticated wall-following behavior, and signal processing for averaging the light measurement being performed in parallel.

11.4   Behavior Trees

A behavior tree (Colledanchise and Ögren 2018) provides structure for hierarchically organizing the decision-making flow of a system that makes many of the considerations that need to be explicitly coded in an FSM implicit instead. The leaves of a behavior tree are "action nodes" that can represent actual discrete behaviors, such as "close gripper" or "find block." The root and internal nodes of the behavior tree are made up of "utility nodes" that guide the path of traversal through the tree. What equates to manual addition and removal of state transitions in an FSM can often be accomplished in a behavior tree by simply changing the type of an utility node from one to another. Another powerful aspect of this abstraction is that behavior trees specifying complex behaviors, such as "navigate to kitchen," can be encapsulated within a single node of another tree.

11.4.1   Node Definition and Status

In a traditional implementation, the nodes within a behavior tree can return any of three statuses when queried: "Success," "Failure," or "Running." The incorporation of a "Running" status allows the behavior tree to use behaviors that operate over longer time periods, such as a block-picking behavior that persists over multiple control cycles of a robot's main processing loop, including time required to plan the end-effector's path, the time required to physically move the robot to the destination, and the time required to close the gripper. In this example, the node might return "Failure" if any of the individual behaviors didn't work or if the end-effector didn't successfully grasp the block by the end of the behavior, and "Success" otherwise. Thus, each node in the behavior tree needs a rigidly defined notion of "Success" or "Failure" that can be propagated throughout the behavior tree, informing which sequence of behaviors is executed to achieve the desired result.

Unlike the FSM formalism that didn't incorporate an explicit notion of time, the "Running" status allows nodes to operate using the information that their child nodes may take variable amounts of time, with each discrete unit of time defined as a *tick*. This design choice simplifies the specification of control flow and dramatically reduces the number of explicit transitions that are needed to model a system. Suffice to say, for a robot with a 100 ms control loop, many of the discrete behaviors that a programmer would be interested in (such as turning 180 degrees or moving forward one meter) are likely to require more than a single program cycle and will have action nodes that run for multiple ticks.

Nodes may also be parameterized, allowing for information computed from one node to be passed on and used in a subsequent node. Consider building a behavior tree for sorting blocks on a table into bins by color. One way to organize this behavior tree is repeating the sequence of behaviors—"find block," "pick block," "get block color," "place block in bin"—until no blocks remain. In this case, the behaviors "get block color" and "place block in bin" are connected, since the color of the block will determine which bin it should be

Table 11.1
Common behavior tree nodes and their symbols.

Node class	Node type	Symbol
Composite	Sequence	$\rightarrow$
Composite	Selector/fallback	?
Composite	Parallel	$\rightarrow$ $\rightarrow$
Decorator	Decorator	$\Diamond$
Action	Action	Text

placed in. This potential for interaction between nodes allows for powerful expressiveness of complex behaviors.

11.4.2 Node Types

Within a behavior tree, nodes can generally be classified based on their connectivity (i.e., do they have children, and if so, how many?) and function (i.e., is this a utility node that determines control flow, or is it an action node executing the action itself?). The three primary node types are *composite*, *decorator*, and *action* and are summarized in Table 11.1.

Composite nodes have one or more children and are responsible for regulating the control flow. Three important examples of composite nodes are the *sequence node, selector/fallback node*, and *parallel node*. A sequence node executes each of its child nodes in order, returning "Failure" if a single one fails and "Success" after all have finished successfully. A selector (or fallback) node executes each of its child nodes in order but returns "Success" once a single child node succeeds, only returning "Failure" if all child nodes have failed. Sequence nodes can be thought of as analogous to an *AND* conditional statement, while selector nodes are similar to an *OR* conditional statement. A parallel node has $N > 1$ children and attempts to execute its child nodes in parallel, returning "Success" if M or more children succeed and "Failure" if more than $(N - M)$ children fail, for any choice of $M \leq N$.

Decorator nodes have exactly one child node and perform transformations on the child node's outputs back to its parents. An example of a simple decorator node is the *inverter*, a node that inverts the return status of its child, effectively producing a *NOT* operation: If the child node returns "Success," then the decorator returns "Failure," and vice versa. Another useful decorator node is one that returns "Success" when its child returns a status of either "Success" or "Failure," allowing for the inclusion of action nodes where success is

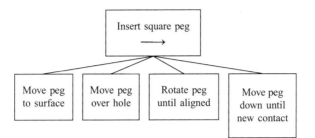

Figure 11.6
Behavior tree for a peg insertion task. A sequence node triggers the execution of movement actions that align a peg over a hole and lower it until the gripper makes contact with the surface.

not critical for the behavior. Decorator nodes can also be designed to repeat the execution of its child—for instance until it returns a "Success" status, until it returns a "Failure" state, or endlessly (which is typically placed as the tree's root node to ensure continuous operation).

Action nodes have zero child nodes and represent the execution of a discrete behavior. These nodes can use input parameters, return output values, and generally have any amount of complexity that the designer desires to program within them. Crucially, an entire behavior tree can be treated as a single action node, allowing for the composition of multiple behavior trees to build arbitrarily complex behaviors.

11.4.3   Behavior Tree Execution

For each unit of time (e.g., control cycle) that passes, a preorder tree traversal occurs where nodes are recursively visited and evaluated left-to-right, commonly described as *propagating a tick* signal through the tree. In doing so, each parent node calls on its child nodes in order to retrieve their status. If a child node returns "Success," the parent node will move on to its next child node. If a child node returns a status of "Running," then the parent node will return "Running" without moving on to the next child node unless it permits running multiple child nodes in parallel. If a child node returns a status of "Failure," its behavior will depend on the type of node the parent is—for example, returning "Failure" if the parent is a sequence node or moving on to the next child node if it is a selector node.

Consider the example of a robotic manipulator inserting a peg into a hole in figure 11.6. The first tick through the tree will trigger the *Move peg to surface* action. Subsequent ticks will be absorbed into this action until it returns a status other than "Running," at which point the next action will be triggered and the same absorption of ticks will occur in the new action being executed. If any single action node returns a status of "Failure," the entire behavior will result in a "Failure" status. A slightly more complex behavior tree is demonstrated with the *Pick square peg* behavior shown in figure 11.7, which allows the robot to check whether or not it is holding a peg already, only moving its gripper for the pick action if this check fails.

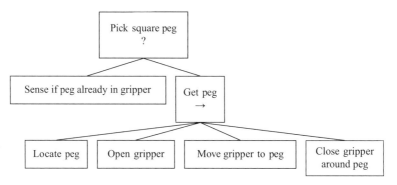

Figure 11.7
Behavior tree for a peg picking task. First, the selector node will check if a peg is already in the robot's gripper. If it is not, the first action node will fail and the *Get peg* behavior will execute, returning "Success" only if the gripper is successfully closed around the peg.

11.4.4 Implementation

As the execution of a behavior tree is fundamentally a tree traversal, a tree is the ideal data structure to store composite, decorator, and action nodes. Since the mechanics of these nodes largely remain the same, nodes are usually implemented as classes (in an object-oriented programming sense), which need to be inherited, modified, and instantiated by the programmer. Programming a complex robotic system therefore starts with defining and implementing the basic action nodes and then recombining them with appropriate composition nodes and decorators until the robot performs as expected.

11.5 Mission Planning

So far, we have seen how reactive behaviors can be composed into more complex programs using finite state machines and behavior trees. Although behavior trees facilitate dealing with the exploding number of possible state transitions by making them implicit, the programmer still needs to define the entire program flow. Consider again a pick-and-place task. This time, we will not simply grasp a new item in case the object falls out of the hand, but try to find it on the table and try to pick it up from there. In a more elaborate version, we might also go on and search for the object on the floor if it cannot be found on the table. But why not have the robot replace an object from a warehouse if it cannot be found, or even mail-order a new version? Obviously, it is very cumbersome to foresee all these eventualities when programming a robot. We therefore need a framework to make it easier to compose behaviors in real time. This is where *mission planning* excels.

An example of mission planning is described in Saito et al. (2011), where a robot is tasked to deliver a sandwich. The robot initially moves to a fridge, opens it, and looks for a sandwich there, then decides to take the elevator to the sandwich store in the basement of the University of Tokyo's engineering tower. Here, the robot is not only piecing together behaviors as it goes, but it is also using what is known as "semantic planning" to select the right actions, exploiting databases of commonsense knowledge in textual form. How to represent such knowledge in an efficient and general manner is an active research topic in robotics and artificial intelligence and goes beyond the scope of this introductory book. However, we describe here the basic algorithms that will allow you to compose complex behaviors at run-time, thereby generating much more complex robot responses than could ever be accomplished using hand-coding.

11.5.1   The General Problem Solver and STRIPS

One of the first planning frameworks was introduced more than 60 years ago as the "General Problem Solver" (Newell, Shaw, and Simon 1959, 64), an idea that was popularized, refined, and actually demonstrated on real robots later on as the "STanford Research Institute Problem Solver" (STRIPS) (Fikes and Nilsson 1971). In STRIPS, a robotic problem is composed of the following elements:

1. A set of symbols that represent the *initial state*
2. A set of symbols that represent the desired *goal state*
3. A set of *actions*, each with a set of *preconditions* and a set of *postconditions*

An action's preconditions are a set of symbols that need to be part of the current state for the action to execute. An action's postconditions are the set of symbols the action creates or deletes, thereby affecting the state. In a nutshell, a STRIPS planner will work backward from a desired goal state, find actions that have equivalent postconditions, and then recursively try to satisfy these actions' preconditions.

For a robot to be able to get you a sandwich, a suitable goal state could be ROBOT HAS SANDWICH=*true*; some possible actions could be PICK SANDWICH FROM FRIDGE and OPEN FRIDGE DOOR; and initial states might be SANDWICH IN FRIDGE=*true* and FRIDGE DOOR CLOSED=*true*. An action OPEN FRIDGE DOOR would then require FRIDGE DOOR CLOSED=*true* as a precondition and lead to FRIDGE DOOR CLOSED=*false* as a postcondition. The action PICK SANDWICH FROM FRIDGE would require SANDWICH IN FRIDGE=*true* and FRIDGE DOOR CLOSED=*false* as preconditions and result in ROBOT HAS SANDWICH=*true*. A planner can now work backward from the desired state to identify appropriate actions and then satisfy their preconditions recursively. In this case, the planner will discover the action PICK SANDWICH FROM FRIDGE and then identify OPEN FRIDGE DOOR to satisfy the precondition FRIDGE DOOR CLOSED=*false*.

Formally, an instance of a STRIPS problem is a quadruple $\langle P, O, I, G \rangle$:

• P is a set of propositional variables that can be either true or false and that exhaustively describe the world the robot operates in.

• O is a set of operators, each itself a quadruple $\langle \alpha, \beta, \gamma, \delta \rangle$ whose entries determine the set of conditions that need to be true (α) and that need to be false (β) for the action to take place, and a set of conditions that will be true (γ) and that will be false (δ) if the action is successful.

• I is a set of conditions $I \subset P$ that are initially true and define the initial state, while all other conditions are initially false.

• G is a tuple $\langle N, M \rangle$ in which N is a set of conditions that need to be true and M is a set of conditions that need to be false.

Formalizing the sandwich example from above using this framework is left to the reader as an exercise. It becomes quickly clear that the devil is in the details here. For example, we have assumed that the positions of the robot are resolved by the action themselves. In practice, the STRIPS plan would also require additional preconditions on the robot's location (e.g., ROBOT AT FRIDGE=*true*, which would then be resolved by the planner). An observant reader might have also noticed that great care needs to be taken in determining which variables an action actually affects and specifying the precisely desired goal state. For example, the plan as described above will lead to a scenario in which the fridge door remains open.

A common extension in a STRIPS instance is to parameterize locations and objects. In this case, ROBOT AT FRIDGE would become ROBOT AT X. Values for "X" can then be substituted at run-time, for example, when evaluating the preconditions of OPEN FRIDGE. Similarly, a STRIPS plan might be formulated to satisfy hunger, substituting the sandwich with other victuals. Managing these different categories, their contexts, and trade-offs between qualities of different outcomes becomes quickly challenging and is an ongoing subject of research.

Other challenges with STRIPS planning are exogenous events that change the environment state, such as a draft that closes the fridge door after the robot has opened it or probabilistic operators that might lead to different outcomes for an operator depending on chance. These are situations that are well covered by the behavior tree (BT) framework, making the combination of BT and STRIPS planning particularly compelling (see chapter 7 of also Colledanchise and Ögren 2018).

Take-Home Lessons

1. Writing a robotic program is fundamentally different from regular computer programs as the program flow needs to be coordinated with the actual physics of the world taking its course.

2. Discrete states are an abstraction of the physical world, and more complex behaviors lead to exponential growth in the number of states and transitions between them.

3. Multiple programming paradigms are available that make managing large number of states and possible transitions between them more manageable, but they require an increasing amount of software and thereby increase computational infrastructure.

Exercises

1. A differential-wheel robot has three downward-facing light sensors at its tip. The sensors are spaced such that the robot can detect a black line on a white ground. Derive the equations for a line-following robot using the Braitenberg formalism.

2. Derive a control scheme that combines line following and obstacle avoidance. Discuss your choices, assuming that the robot has to avoid obstacles at all cost.

3. Use a robotic simulator of your choice to implement basic phototaxis and obstacle avoidance.

4. Use a robotic simulator of your choice to implement basic wall-following behavior.

5. Implement a light-following robot in a simulator of your choice and manually control it toward the light. Train a neural network for a Braitenberg controller using this data.

6. Implement a simple finite state machine that combines obstacle avoidance, phototaxis, and wall following and is able to escape from a U-shaped obstacle.

7. Consider an FSM that performs phototaxis until encountering an obstacle, then performs wall following for 10 time-steps. Draw an appropriate finite state machine that implements this behavior. How many states do you need?

8. A robot runs at a 100 ms loop time. Performing sensor readings takes 3 ms, odometry computations 15 ms, and executing logic takes 30 ms on average. Which of these operations is likely to fail if the task logic takes 80 ms?

9. Formulate both a finite state machine and a behavior tree for the game "Ratslife," label each state and conditional transition, and compare the two representations.

10. Construct a behavior tree that enables a single robot manipulator arm to sort red, green, and blue blocks on a table into bins by color.

11. Construct a behavior tree that enables a bimanual (two manipulator arm) robot to sort red, green, and blue blocks on a table into bins by color with both arms.

12. Formally describe a STRIPS instance for a robotic sandwich retrieval problem in which the fridge door is closed after the robot has retrieved a sandwich.

12 Mapping

Mapping is the process of building representations of the environment for either downstream consumption by autonomy algorithms or for informing humans. Maps inform decision-making for planning and control algorithms by, for example, providing information on surfaces and obstacles, objects with which the robot can interact, or topological information such as how rooms are connected to each other. If maps of an environment are already provided, robots can build plans over them without having to build a map themselves; indeed, they can even localize themselves within these maps simply by collecting information in situ and referencing that information against prior maps. Mapping also provides humans with an appreciation for what the robot *sees*, thereby guiding designers or operators of robots with information about what is possible or what information is available to the robot. Therefore, map information for both the purposes of autonomy and design is a critical linchpin for operating in real environments.

When one considers the quantities of interest to be collected from an environment, there are two distinct classes: metric (e.g., the physical extents of an environment) and semantic (e.g., the type of room one is in, or objects of interest within it). These two categories can inform one another. For instance, a door is typically of a particular height or width, but the inference required to resolve information in these two classes is remarkably different. Metric information is frequently resolved using geometric techniques with which we have become familiar in chapter 7, whereas semantic information is typically obtained through machine learning techniques. For the purposes of this text, we focus almost exclusively on the problem of metric mapping because of the prevailing importance of this information as it relates to robotic planners and localization, but semantic mapping is a rich field that is increasingly important for developing robotic autonomy. Greater distinctions between these types of maps will be discussed in section 12.1.

With respect to metric mapping, range sensors have emerged as one of the most effective sensors to make robots autonomous. Range data can be collected into "scans" from a sensor, each of which consists of a list of points that we term a point cloud. Point clouds make the construction of a three-dimensional (3D) model of the robot's environment straightforward; measurements on the environment are inherently metric and require no "front-end" processing, as images from a camera might require to extract such information. Furthermore, sensors that output point cloud data are both quite accurate and increasingly common on robotic platforms: the Velodyne 3D automotive lidar sensor that combines 64 scanning lasers into one package was key in mastering the DARPA Grand Challenge, and no team operated without lidars as part of their solution in the recent DARPA Subterranean Challenge. So, for both wide-area and close-corridor operation, 3D lidar has become a standard. There is one hitch with lidars: Most of them are built out of rotating laser arrays, which means that a moving sensor will collect information from the environment at different rotation angles, thereby *aliasing* the lidar measurements with the motion of the sensor. This motion aliasing can be removed from the lidar data but is negligible if the motion of the sensor is slow enough. However, there are sensors that provide range data without this limitation, such as RGB-D (color plus depth) cameras. Furthermore, 3D range data has become even more important in robotics with the advent of cheap RGB-D cameras that are a tenth of the price of the cheapest two-dimensional (2D) laser scanner. In this chapter we largely ignore the source of range data and instead focus on algorithms that operate over these data.

The mapping problem itself also ranges from trivial (when localization is perfect) to arbitrarily complex, performing the equivalent of cartography in which the tasks of localizing oneself and mapping the environment are closely intertwined. While the problem of "simultaneous localization and mapping" will be introduced in chapter 17, section 12.2 describes one of the key algorithms that is used to infer relative pose between consecutive measurements by a process known as scan matching.

Point cloud data readily admits the identification of features such as lines and planes using random sample and consensus (RANSAC). These features can be consumed by extended Kalman filter (EKF)-based localization but can also be used for improving odometry, loop-closure detection, and mapping. Point cloud data can also be probabilistically fused into voxel occupancy grids and dense surface representations, which inform planning and design. The goals of this chapter are as follows:

• Introduce the iterative closest point (ICP) algorithm for matching point clouds as an example of sparse mapping.

• Show how ICP can be improved by providing initial guesses using RANSAC.

• Use point clouds to generate dense maps built through occupancy grids.

• Demonstrate RGB-D mapping, another dense mapping technique that results in surface representations.

12.1   Map Representations

In order to plan a path, we need to represent the environment digitally. We differentiate between two complementary approaches: discrete and continuous approximations. In a discrete approximation, a map is sub-divided into sections of equal (e.g., a grid or hexagonal map) or differing sizes (e.g., rooms in a building). The latter maps are also known as *topological maps* or *graph-based maps*. Discrete maps lend themselves well to a graph representation. Here, every region of the map corresponds to a vertex (also known as a "node"), which is connected by edges if a robot can navigate from one vertex to the other. For example, a road map is a topological map with intersections as vertices and roads as edges, labeled with their length (figure 13.2). Computationally, a graph might be stored as an adjacency or incidence list/matrix. A continuous approximation requires the definition of inner (obstacles) and outer boundaries, typically in the form of a polygon, whereas paths can be encoded as sequences of points defined by real numbers. Despite the memory advantages of a continuous representation, discrete maps are the dominant representation in robotics.

There is no one correct choice for choosing a map representation, and each application might require a different solution that could use a combination of different map types.

Discrete and continuous representations are often matched together in clever ways. For example, road maps for GPS systems are stored as topological maps that store the GPS coordinates of every vertex, but they might also contain overlays of aerial and street photography. These different maps are then used at different stages of the path planning stage.

12.2   Iterative Closest Point for Sparse Mapping

In its simplest form, a map can be created from slices of 2D range data such as obtained from a laser scanner. In the absence of a precise estimate of motion between two measurements— for example, as provided by odometry or data from an inertial measurement unit (IMU)— the challenge is to associate subsequent scans.

A standard solution to this problem is known as the *iterative closest point* (ICP) algorithm. It was presented in the early 1990s for registration of 3D range data to computer-aided design (CAD) models of objects. A more in-depth overview of what is described here is given in Rusinkiewicz and Levoy (2001). The key problem can be reduced to finding the best transformation that minimizes the distance between two sets of measurements.

In robotics, ICP found an application to match scans from 2D laser range scanners. For example, the transformation that minimizes the error between two consecutive snapshots of the environment is proportional to the motion of the robot. This is a hard problem because it is unclear which points in the two consecutive snapshots are "pairs," which of the points are outliers (due to noisy sensors), and which points need to be discarded as not all points overlap in both snapshots. Stitching a series of snapshots together theoretically allows us to create a 2D map of the environment. This is difficult, however, as the error between every snapshot—similar to odometry—accumulates. The ICP algorithm also works in 3D where it allows us to infer the change in a six-dimensional (6D) pose of a camera and the creation of 3D maps. In addition, ICP has proved useful for identifying objects from a database of

3D objects. Furthermore, the ICP algorithm can be used to stitch consecutive range images together to create a 3D map of the environment (Henry et al. 2010).

Before providing a solution to the mapping problem, we will focus on the ICP algorithm to match two consecutive frames. Variants of the ICP algorithm can be broken down into six consecutive steps:

1. Selection of points in one or both meshes or point clouds
2. Matching/pairing these points to samples in the other point cloud/mesh
3. Weighting the corresponding pairs
4. Rejecting certain pairs
5. Assigning an error metric based on the point pairs
6. Minimizing the error metric

Depending on the number of points generated by the range sensor, it might make sense to use only a few selected points to calculate the optimal transformation between two point clouds and then test this transformation on all points. Depending on the source of the data, it also turns out that some points are more suitable than others because it is easier to identify matches for them. This is the case for RGB-D data, where scale-invariant feature transform (SIFT) features have been used successfully. This is also the case for planar objects with grooves, where sampling should ensure that angles of normal vectors of sampling points are broadly distributed. Which method to use is therefore strongly dependent on the kind of data being used and should be considered for each specific problem.

Matching points
The key step in ICP is to match a point to its corresponding point in a different measurement. For example, a laser scanner hits a certain point at a wall with its 67th ray. After the scanner has been moved by 10 centimeters (cm), the closest hit on the wall to this point might have been by the third ray of the laser. Here, it is actually very unlikely that the laser hits the exact same point on the wall twice, therefore introducing a nonzero error even for an optimal pairing. Prominent methods involve finding the closest point in the other point cloud or finding the intersection of the source point's normal with the destination surface (for matching point clouds to meshes). More recently, SIFT has made it possible to match points based on their visual appearance. Similar to sorting through SIFT features, finding the closest matching point can be accelerated by representing the point cloud in a k-d tree.

Weighting of pairs
As some pairs are better matches than others, weighting them in some principled way can drastically improve the quality of the resulting transformation. One approach is to give more weight to points that have smaller distances from each other. Another approach is to take into account the color of the point (in RGB-D images) or use the distance of their SIFT features (weighting pairs with low distances higher than pairs with high distances). Finally,

expected noise can be used to weight pairings. For example, the estimates made by a laser scanner are much more faithful when taken orthogonally to a plane than when taken at a steep angle.

Rejecting of pairs

A key problem in ICP are outliers either from sensor noise or simply from incomplete overlap between two consecutive measurement frames. A common approach to deal with this problem is to reject pairings when one of the points lies on a boundary of the point cloud, as these points are likely to match with points in nonoverlapping regions. As a function of the underlying data, it might also make sense to reject pairings with too high of a distance. This is a threshold-based equivalent to distance-based weighting as described above.

Error metric and minimization algorithm

After points have been selected and matched and pairs have been weighted and rejected, the match between two point clouds needs to be expressed by a suitable error metric that will then need to be minimized. One straightforward approach is to consider the sum of squared distances between each pair. This formulation can often be solved analytically. Let

$$A = \{a_1, \ldots, a_n\} \tag{12.1}$$

$$B = \{b_1, \ldots, b_n\} \tag{12.2}$$

be point clouds in $\mathbb{R}^n$. The goal is now to find a vector $t \in \mathbb{R}^n$ so that an error function $\phi(A + t, B)$ is minimized. In 6D (translation and rotation), an equivalent notation can be found for a transformation (see forward kinematics). An error function for the squared distance is then given by

$$\phi(A + t, B) = \frac{1}{n} \sum_{a \in A} \|a + t - N_B(a + t)\|^2. \tag{12.3}$$

Here $N_B(a + t)$ is a function that provides the nearest neighbor of a translated by b in B. A key problem now is that the actual value of t affects the outcome of the pairing. What might look like a good match initially often turns out not to be the final pairing. A simple numerical approach to this problem is to find t iteratively.

Initially, $t = 0$ and nearest neighbors/pairings are established. We can now calculate a δt that optimizes the least-squares problem based on this matching using any solver available for the optimization problem (for a least-squares solution, δt can be obtained analytically by solving for the minimum of the polynomial by setting its derivative to zero). We can then shift all points in A by δt and start over. That is, we calculate new pairings and derive a new δt. We can continue to do this, until the cost function reaches a local minimum.

Instead of formulating the cost function as a "point-to-point" distance, a "point-to-plane" has become popular. Here, the cost function consists of the sum of squared distances from each source point to the plane that contains the destination point and is oriented perpendicular to the destination normal. This particularly makes sense when matching a point cloud to a mesh/CAD model of an object. In this case, there are no analytical solutions to finding the optimal transformation, but any optimization method (such as Levenberg-Marquardt) can be used.

12.3   Octomap: Dense Mapping of Voxels

For mapping obstacles, the most common map is the *occupancy grid map*. In a grid map, the environment is discretized into *voxels* of arbitrary resolution (e.g., 1 cm × 1 cm), upon which obstacles are marked. In a probabilistic occupancy grid, grid cells can also be marked with the probability that they contain an obstacle. This is particularly important when the position of the robot that senses an obstacle is uncertain. Disadvantages of grid maps are their large memory requirements as well as the computational time required to traverse data structures with large numbers of vertices. A solution is storing the grid map as a *k-d tree*. A k-d tree recursively breaks the environment into *k* pieces, subject to a subdivision rule (e.g., only subdivide a space if it is between 5 to 95 percent occupied). For $k = 4$, an area that fits the subdividing criteria would be subdivided into four pieces. Each of these pieces can again be subdivided into four pieces and so on, until the maximum allowable resolution is reached or the subdivision criteria no longer applies. These pieces can be stored in a graph with each vertex having four children, corresponding to the four pieces the space represented by the vertex is broken into, unless it is a leaf of the tree. This data structure is attractive because not all vertices need to be broken down to the smallest possible resolution. Instead, only areas that contain obstacles need to be subdivided. A grid map containing obstacles and the corresponding k-d tree are shown in figure 12.1. To capture 3D data, this representation can be extended to an 8-d tree, also known as an *octree*.

The values of each entry in the k-d tree is the probability that the particularly specified voxel is occupied. Note that this probability may be calculated through any number of sensor models, such as absolute thresholding or using a probabilistic field-of-view sensor model. Absolute thresholding determines that a voxel is occupied if the absolute count of points measured by the sensor within a voxel is greater than a threshold. An improvement on this is to use a probabilistic model wherein a false-positive incidence rate and a false-negative incidence rate are used to calculate the probability that a sensor's measurements indicate that a particular voxel is filled. Either way, these techniques result in a map that is probabilistically fused over a sequence of measurements to indicate filled and unfilled space in the form of a volumetric map.

12.4   RGB-D Mapping: Dense Mapping of Surfaces

While occupancy grid mapping using a technique such as octomap is efficient for planning, there are some drawbacks. First, it can be immediately observed that the map voxels are of a fixed resolution and fundamentally cannot resolve small obstacles at any smaller

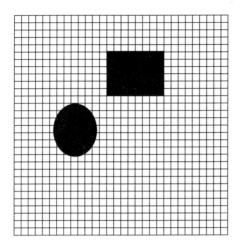
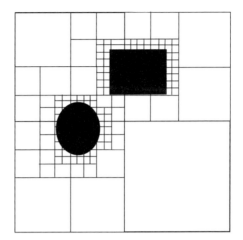

Figure 12.1
A grid map and its corresponding quadtree (k-d tree).

scale than that of the voxels; that is, small obstacles will appear larger. Furthermore, high-frequency information within the voxel that may be relevant to planning, such as surface curvatures, are unresolvable. However, this information seems out of reach for any voxelized representation of the environment. The resolution to this paradox is to populate the value of voxels not with probability of a voxel's being occupied, but rather the *most probable distance* to the nearest surface. If, for a particular range scan, a surface lies beyond a certain voxel, that distance is positive, and if a surface lies in front of a certain voxel, that distance is negative. See figure 12.2 for an example of this mathematical construction, which is known as a *signed distance field* (SDF). The SDF is generated by following the distance along a ray to the surface, entering that value into the voxel, and incrementally probabilistically updating the values in the voxel as frames are consumed from the depth channel. Note that the SDF provides an implicit representation of a surface, as can be seen in figure 12.2. In this figure the notion of "truncation" is also motivated, wherein voxels that would have a value above a certain threshold known as a truncation distance are left unfilled; this is a frugal use of memory and results in a speed-up for reconstruction algorithms. It also is required to avoid surfaces interfering with one another. An SDF that is pruned in this way is known as a "truncated SDF" (TSDF).

The TSDF can naturally represent multiscale obstacles and has an added benefit for planning algorithms: It provides the distance to the nearest obstacle at the same time! This is helpful since distance to obstacles can be used as a risk metric in planning algorithms (i.e., it is often advantageous to maintain maximal distance from obstacles over a trajectory, which can be easily obtained from the TSDF). There are two significant downsides of this technique, however. First, it requires highly accurate pose information, frequently meaning ICP

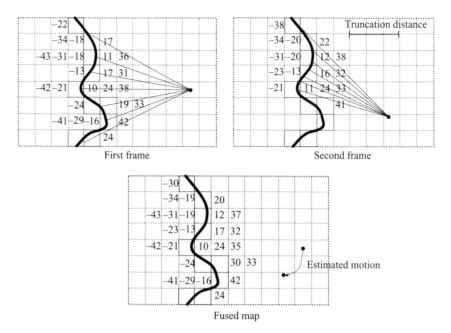

Figure 12.2
Schematic of the generation of a TSDF based on 2D range data from a sensor. Voxels are only populated with distances that lie within the "truncation distance."

must be performed for each scan from the sensor. Second is that implicit representation of surfaces do not admit straightforward visualizations of the 3D maps. To accomplish this, a renderer is required to operate on the TSDFs, which results in maps such as the one shown in figure 12.3, which is created using the method of Whelan et al. (2013). The resulting visualizations can be of a surprisingly high resolution even over coarse voxelization of the environment. Together with RGB information, it is possible to create complete 3D walk-throughs of an environment.

The problem with using ICP continuously to generate these maps is that errors in each transformation propagate into the map generation process in the form of map drift. Here, the simultaneous localization and mapping (SLAM) algorithm (chapter 17) can be used to correct previous errors once a loop closure is detected, but the update of TSDFs on the trigger of a loop closure requires both the continuous retaining and global reprocessing of all data used to generate the TSDFs that would be affected by the loop closure, which is a high price to pay.

As ICP only works when both point clouds are already closely aligned, which might not be the case for a fast-moving robot with a relatively noisy sensor (the Xbox Kinect has an error of 3 cm for a few meters of range versus millimeters in laser range scanners), RGB-D mapping uses RANSAC to find an initial transformation. Here, RANSAC works as for line

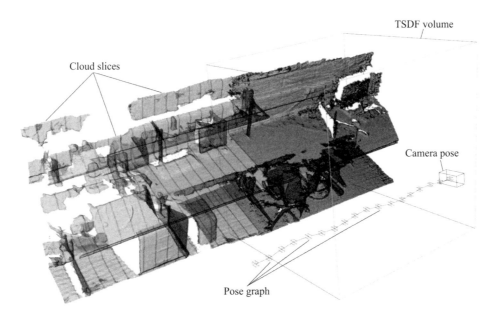

Figure 12.3
Fused point cloud data from a walk-through of an office environment using "Kintinuous." *Source*: Picture courtesy of John Leonard.

fitting: It keeps guessing possible transformations for three pairs of SIFT feature points and then counts the number of inliers when matching the two point clouds, one of which is being transformed using the random guess.

Take-Home Lessons

1. The challenge in mapping an environment originates from the uncertainty in both localization and sensing.

2. Techniques that are used to overcome uncertainty in localization and sensing can in turn be used to increase confidence in the former. For example, given robust features such as corners or walls, ICP results can be used to improve odometry estimates.

3. In the absence of reliable localization, the mapping problem turns into the simultaneous localization and mapping problem that is addressed in chapter 17.

Exercises

1. Simulate a lidar sensor in a simulator of your choice. Devise a grid-map structure that will allow you to draw the robot's position. Use the constant angular offset

of your lidar sensor and the pose of the robot to compute map coordinates for each reading.

2. Run a simulated robot in an obstacle course and record a map using your simulated lidar. Implement the ICP algorithm described above to estimate the translation between consecutive scans and compare them with your odometry estimate.

3. Use ICP to improve the robot's state estimate for different settings of wheel-slip in your simulation.

13 Path Planning

Path planning allows autonomous mobile robots and manipulators to find a path to move between two points. A *path* is a set of poses from a start configuration to an end configuration that respects a set of specifications (e.g., avoiding obstacles for a mobile base or respecting a specific force profile at the end-effector of a manipulator). It differs from the concept of *trajectory*, which is the execution of a path over time. Depending on the choice of the planning algorithm, a path could satisfy various degrees of optimality with respect to criteria such as minimizing path length, minimizing turns, or minimizing the amount of braking. Algorithms to find a shortest path are important not only for robotics applications but also in network routing, video games, and understanding protein folding.

Path planning requires a suitable representation of the environment, such as a map introduced in chapter 12, and a perceptual understanding of the robot's location with respect to such representation We will assume for now that the robot is able to localize itself, is equipped with a map, and is capable of avoiding temporary obstacles on its way. The goals of this chapter are to:

- Introduce the concept of "configuration space" for planning.
- Understand the difference between graph-based and sampling-based planning algorithms.
- Explain basic path algorithms such as Dijkstra, A*, and RRT.
- Understand variations of the path planning problem such as coverage path planning.

13.1 The Configuration Space

In the vast majority of path planning algorithms, the robot is treated as a point-mass element with no volume. In order for a path to be executed on the robot, it is important to take into account the physical embodiment of the robot and its nonzero volumetric occupancy, which complicates the path planning process. It is possible for the robot to be reduced to a point-mass while growing all obstacles by its radius. This works for a circular robot. This can be generalized for robots of any shape by growing each obstacle by the length of the longest extension of the robot from its center. This representation is known as *configuration space*

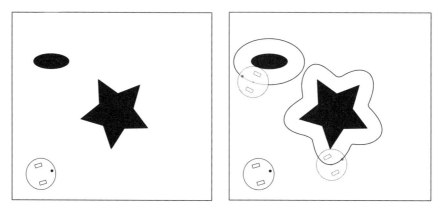

Figure 13.1
A map with obstacles and its representation in configuration space, which can be obtained by growing each obstacle by the robot's extension.

because it reduces the representation of the robot to its controllable degrees of freedom (e.g., its x and y coordinates in the plane for a robot capable of planar translation). An example is shown in figure 13.1. The configuration space can now either be used as a basis for a grid map or a continuous representation.

13.2   Graph-Based Planning Algorithms

The problem to find a "shortest" path from one vertex to another through a connected graph is of interest for multiple domains, most prominently network routing, where it is used to find an optimal route for an internet data packet. The term "shortest" here is defined as the minimum cumulative edge cost, which could be physical distance (in a robotic application), delay (in a networking application), or any other metric that is relevant for the task. An example graph with arbitrary edge lengths is shown in figure 13.2.

13.2.1   Dijkstra's Algorithm

One of the earliest and simplest algorithms for path planning is Dijkstra's algorithm (Dijkstra 1959). Given a graph, Dijkstra is an iterative process where, starting from the "start" vertex, the algorithm marks all its direct neighbors with the cost to reach them. It then proceeds to inspect the neighboring vertex with the lowest cost and all its adjacent vertices and marks them with the cost to get to them via the vertex under consideration. If the cost turns out to be lower, the cost is updated accordingly. Once all neighbors of a vertex have been checked, the algorithm proceeds to the vertex with the next lowest cost. Once the algorithm reaches the goal vertex and there is no vertex with a lower cost to the goal, it terminates and the robot can follow the edges pointing toward the lowest edge cost.

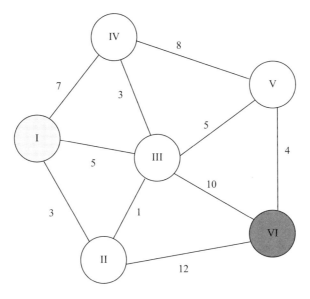

Figure 13.2
A generic path planning problem from vertex I to vertex VI. The shortest path is I-II-III-V-VI, and has length of 13.

In the example in figure 13.2, Dijkstra would first mark nodes II, III, and IV with cost 3, 5, and 7, respectively. It would then continue to explore all edges of node II, which so far has the lowest cost. This would lead to the discovery that node III can actually be reached in $3 + 1 < 5$ steps, and node III would therefore be relabeled with cost 4. In order to completely evaluate node II, Dijkstra needs to evaluate the remaining edges before moving on and label node VI with $3 + 12 = 15$. The node with the lowest cost is now node III (cost of 4). We can now relabel node VI with 14, which is smaller than 15, and label node V with $4 + 5 = 9$, whereas node IV remains at $4 + 3 = 7$. Although we have already found two paths to the goal, one of which is better than the other, we cannot stop because there are still other nodes with unexplored edges and overall cost lower than 14. Indeed, continuing to explore from node V leads to a shortest path I-II-III-V-VI of cost 13, with no remaining nodes to explore.

As Dijkstra would not stop until there is no node with lower cost than the current cost to the goal, we can be sure that a shortest path will be found if it exists. We can therefore say that Dijkstra is both *complete* and optimal.

As Dijkstra will always explore nodes with the least overall cost first, exploration of the environment resembles a wave front originating from the start vertex, eventually arriving at the goal. This is of course highly inefficient, especially if Dijkstra is exploring nodes away from the goal. As an example, if we were to add a couple of nodes to the left of node I in figure 13.2, Dijkstra would explore all these nodes until their cost exceeds the lowest found

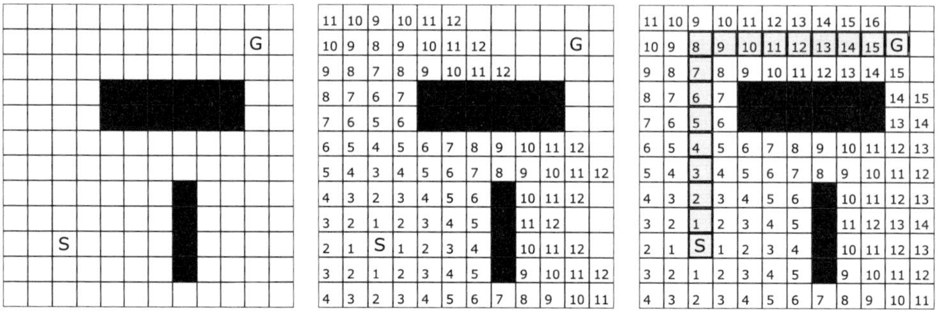

Figure 13.3
Using the Dijkstra's algorithm to find a shortest path from "S" to "G," assuming the robot can only travel laterally (not diagonally) with cost one per grid cell. Note the few number of cells that remain unexplored once the shortest path (gray) is found, as Dijkstra would always consider a cell with the lowest path cost first.

for the goal. This can also be seen when observing Dijkstra's algorithm on a grid, as shown in figure 13.3.

Note that the grid can be reduced to a graph in which each vertex, except those at the borders, have four or eight neighbors.

13.2.2 A*

Instead of exploring in all directions, knowledge of an approximate direction of exploration to reach the goal may help in avoiding the exploration of nodes that are not needed to succeed in the task. As humans, we can easily interpret the task in figure 13.3 and understand that most states in the top-left and bottom-right corner should not be explored if we want to find a solution in a short amount of time. Such knowledge may be encoded in the search algorithm using a *heuristic function* (i.e., an informed guess or estimate of sorts). For example, we could give priority to nodes that have a lower estimated distance to the goal than others. For this, we would mark every node not only with the actual distance that it took us to get there (as in Dijkstra's algorithm) but also with the estimated cost to target—for example, by calculating the Euclidean distance or the *Manhattan distance* between the vertex we are looking at and the goal. This algorithm is known as A* (Hart, Nilsson, and Raphael 1968), and it illustrated in figure 13.4 using the Manhattan distance metric. Depending on the environment, A* might accomplish search much faster than Dijkstra's algorithm, and it performs the same in the worst case.

An extension of A* that addresses the problem of expensive re-planning when obstacles appear in the path of the robot is known as D* (Stentz 1994). Unlike A*, D* starts from the goal vertex and has the ability to change the costs of parts of the path that include an obstacle. This allows D* to replan around an obstacle while maintaining most of the already-calculated path.

A* and D* become computationally expensive when either the search space is large (e.g., due to a fine resolution required for the task) or when the dimensions of the search problem

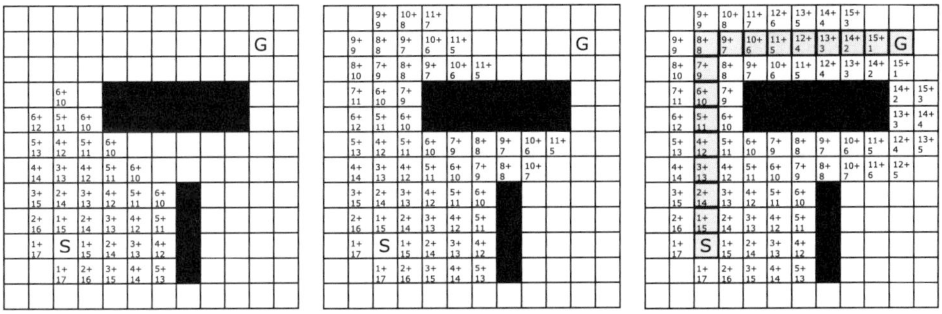

Figure 13.4
Finding a shortest path from "S" to "G" assuming the robot can only travel laterally (not diagonally) with cost one per grid cell using the A* algorithm. Much like Dijkstra, A* evaluates only the cell with the lowest cost but takes an estimate of the remaining distance into account.

are high (e.g., when planning for an arm with multiple degrees of freedom). Solutions to these problems can be provided by sampling-based path planning algorithms.

13.3 Sampling-Based Path Planning

Section 13.2 introduced a series of *complete* algorithms for the path planning problem— that is, algorithms that are guaranteed to (eventually) find a solution if one exists. However, complete algorithms are often infeasible in practice because of a large state space, low available memory, or limited time to execute the algorithm. This is often the case for robots with many degrees of freedom such as arms. Importantly, most algorithms are only *resolution complete*, meaning they are only complete if the choice of environment resolution is fine enough: since the state space needs to be discretized, some solutions might be missed because of such discretization.

Sampling-based planners are an alternative to graph-based planners that evaluate all possible solution and non-complete Jacobian-based inverse kinematic solutions. In sampling-based motion planning, possible paths are generated by random sampling and stored in a tree-like structure until some solution is found or the allotted time expires. As the probability to find a path approaches one when the number of samples goes to infinity, sampling-based path planners are *probabilistic complete*. Prominent examples of sampling-based planners are the rapidly exploring random tree (RRT) (LaValle 1998) and the probabilistic roadmap (PRM) (Kavraki et al. 1996).

An example execution of RRT is shown in figure 13.5; in essence, RRT grows a single tree from a robot's starting point until one of its branches hits a goal. This example illustrates how a sampling-based planner can quickly explore a large portion of space and refine a solution over time. Conversely, probabilistic roadmaps create a tree by randomly sampling points in the state space, testing whether they are collision-free, connecting them

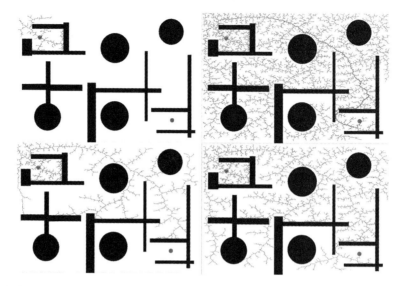

Figure 13.5
Counterclockwise (from top-left): Random exploration of a two-dimensional (2D) search space by randomly sampling points and connecting them to the graph until a feasible path between start and goal is found.

with neighboring points using paths that are achievable subject to the kinematics of the robot, and then using classical graph shortest path algorithms to find shortest paths on the resulting structure. The advantage of PRM is that the map has to be created only once (assuming the environment is not changing) and can then be used for multiple queries. PRM is therefore a *multi-query* path planning algorithm, whereas RRT is known as *single-query* path planning algorithm. Over the years the boundary between these different algorithms has blurred, and single-query and multi-query variants of both RRT and PRM exist. In all, there is no "silver bullet" algorithm or heuristic, and even the choice of their parameters is highly problem-specific. We will therefore limit our discussion on useful heuristics to those that are common to sampling-based planners.

13.3.1 Rapidly Exploring Random Trees

 Let $\mathcal{X}$ be a d–dimensional state space. This can either be the robot's state given in terms of translation and rotations (six dimensions or a subset thereof) or the joint space with one dimension per joint. What representation you chose will determine how to compute whether a point is reachable or not, but it will not affect the algorithm itself.

Let $\mathcal{G} \subset \mathcal{X}$ be a d–dimensional sphere in the state space that is considered to be the goal, `max_dist` the longest permissible edge length, `t` the allowed time, `k` the maximum number of vertices to allow in the tree, and `goal_bias` the percentage of the time the algorithm should try to connect to a goal state. An RRT planner would follow this pseudo-code:

```
Tree=Init(X, G, start, max_dist, t, k, goal_bias);
iteration = 0
WHILE (ElapsedTime() < t AND iteration < k
AND NoGoalFound(Tree,G)) DO:
 iteration = iteration + 1
 IF RandomPercentage() < goal_bias THEN
  q_rand = SampleRandomGoal(G);
 ELSE
  q_rand = SampleRandomState(X);
 ENDIF
 q_nearest = NearestVertex(q_rand)
 q_new = Extend(q_nearest, q_rand, max_dist)
 edge = CreatePath(q_nearest, q_new);
 IF IsAllowablePath(edge) THEN
  Tree.addVertex(q_new);
  Tree.addEdge(edge);
 ENDIF
ENDWHILE
return Tree
```

This process can be iterated as long as time allows and maximum number of vertices or `goal_bias` are optional parameters. RRT is known as an *anytime algorithm*—that is, any user interruption once an initial solution has been found would still provide some kind of solution. Given a suitable distance metric, the path cost can be stored at each node of the tree, allowing one to track the shortest path to goal in case there are multiple vertices in the goal region. There are four key points in this algorithm:

1. Determining the next point `q_rand` to add to the tree (`SampleRandomGoal`, `SampleRandomState`, and `Extend`)

2. Finding out where and how to connect this point to the tree, taking into account the robot kinematics (`NearestVertex`, `CreatePath`)

3. Testing whether this path is suitable (`IsAllowablePath`)—that is, collision-free

4. Smoothing the path (not shown in the algorithm)

Selecting the next best point
A simple approach is to randomly select a point in the state space and connect it to the closest existing point in the tree. Other solutions may assign preference to nodes with few out-degrees (i.e., those without many connections) and choose points in their vicinity in order to facilitate expansion in under-explored regions of the state space. Importantly, both approaches allow one to quickly explore the entire state space; if there are constraints imposed on the robot's path—for example, if the robot needs to hold a cup and therefore is not supposed to rotate its wrist—this dimension can simply be taken out of the state space and fixed at run-time.

Connecting points to the tree

Intuitively, the new point `q_rand` should be connected to the closest point already in the tree or to the goal. This requires iterating over all nodes in the tree and calculating their distance to the candidate point `q_rand`, which is a computationally expensive process; the resulting point `q_nearest` is the one with the shortest distance. The selection of the right data structure for storing the graph in memory may reduce the computational cost to be on average sublinear in the number of vertices.

Importantly, following this method does not guarantee that the shortest path will be found. As an alternative known as RRT* grows the tree in a way that always minimizes the overall path length from the root to every vertex. This is done in two steps. First, only points in the tree within a d−dimensional sphere (on a 2D map, $d = 2$, which is a circle) of fixed radius from `q_rand` are considered, and the point that minimizes the overall path length from the start configuration (rather than simply the shortest distance from `q_rand`) is found. With this step, we can guarantee that the new vertex `q_rand` is connected to the shortest reachable path from the root of the tree. Second, a *rewiring* step occurs where vertices near `q_rand` are evaluated to inspect if an edge between them and `q_rand` would be shorter than the current edge. If this is true and the edge is allowable (i.e., not in collision or outside the physical abilities of the robot), the graph is rewired so that the newfound vertex becomes the new parent of `q_rand`.

Once the nearest vertex is found, the `Extend` function uses the `max_dist` parameter to limit the maximum edge length, replacing `q_rand` with a point `q_new` on the line connecting `q_nearest` and `q_rand` that is `max_dist` away from `q_nearest`. During this step, it is also a good time to take into account the specific kinematics of a robot and its motion capabilities. In the example of a car, a local planner can be used to generate a suitable trajectory that takes into account the orientation of the vehicle at each point in the tree. Using an open-source physics simulation like those developed for computer games also allows one to consider dynamics, including drift. Using such a simulation within a planning framework has demonstrated trajectories that meet the performance of the most skilled operators (Keivan and Sibley 2013).

Collision checking

Efficient algorithms for testing collisions deserve a dedicated section. While the problem is intuitive in configuration-space 2D planning and can be solved using a simple point-in-polygon test (since the robot reduces to a point), this issue becomes more involved for manipulators that are essentially multiple rigid bodies connected together and that may be subject to self-collisions. Conventionally, collision checking for these kind of objects has be achieved by converting them into triangle meshes that can then be tested for intersections. More recently, physics-based computer game engines that provide built-in collision checking are increasingly used. This makes particularly sense when such engines are also used to predict the dynamics of rigid bodies within the `CreatePath` function.

Typically, collision checking takes up to 90 percent of the execution time of a path planning problem; therefore, methods that aim at reducing computational cost are desirable. For example, the "lazy collision evaluation" algorithm differs from standard collision checking in that it does not evaluate every point for a possible collision. Rather, it first finds a suitable path, and only after a path is found does it evaluate every edge for collisions. Segments in collision are deleted and the algorithm continues, but only collision-free segments are maintained.

Once a possible path is found, the sampling space can be reduced to an ellipsoid that bounds the maximal path length. This ellipsoid can be constructed by mounting a wire of the maximum path length between start and goal and pushing it outward with a pen. Intuitively, only points that are contained by this ellipsoidal area can provide a shorter path than the one currently known, so it becomes a waste of time to grow the tree in areas of the state space that are outside this ellipsoid. This approach is particularly effective when running multiple copies of the same planner in parallel and exchanging the shortest paths once they are found (Otte and Correll 2013).

Path smoothing

As path planning randomly samples from discrete and arbitrarily coarse maps, the resulting paths are typically jagged and irregular and far from optimal in practice. This can be drastically improved by path smoothing. One way of doing this is to connect points of the path using splines, polynomial curves, or even trajectory snippets that are known to be feasible for a specific platform. Alternatively, one can also use a model of the actual platform and use a feedback controller such as the one described in section 3.4.2 for mobile robots and section 3.2.2 for manipulators, which will generate a trajectory that the robot can actually drive. When combined with dynamics, this approach is known as *model-predictive control*.

13.4   Planning at Different Length Scales

The reality of performing complex, autonomous behaviors in realistic scenarios is that, in practice, no one map representation and planning algorithm might be sufficient. Planning a route for a car, for example, is a multistep process wherein robot autonomy interleaves with human intelligence. As detailed in figure 13.6, a hierarchy of increasingly granular map representations and path planning algorithm is needed. First, a coarse search is performed over the street network (e.g., using your preferred mapping and navigation app), followed by a more precise planner that determines which lanes to choose and how to navigate roundabouts and intersections; in both these layers of abstraction, graph-based planning algorithms are ideal. Then, a sampling-based algorithm may be used to determine how to actually move the car between lanes and what trajectory to use to avoid obstacles. Finally, such trajectories need to be turned into wheel speeds and steering angles—possibly using some form of feedback control. In figure 13.6, downward-pointing arrows indicate

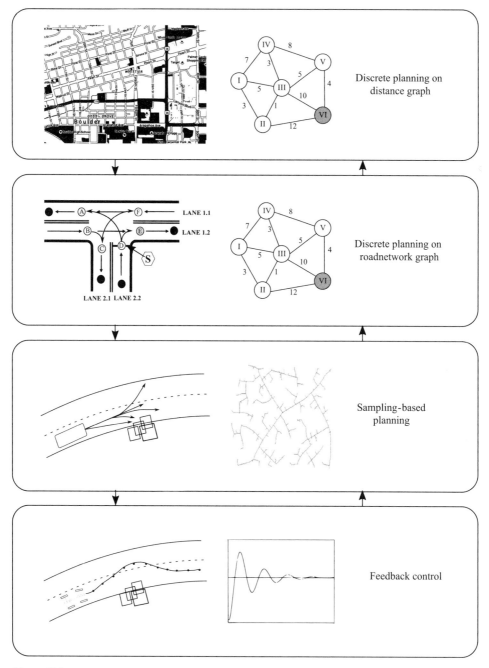

Figure 13.6
Path planning across different length scales, requiring a variety of map representations and planning paradigms.
Arrows indicate information passed between layers.

the input that one planning layer provides to the one below, whereas upward-pointing arrows instead indicate exceptions that cannot be handled at the lower levels. For example, a feedback controller cannot handle obstacles, requiring the sampling-based planning layer to come up with a new trajectory. Should the entire road be blocked, this planner would need to hand off control to the lane-based planner. A similar case can be made for manipulating robots, which also need to combine multiple different representations and controllers to plan and execute trajectories efficiently.

Note that this representation does not include a reasoning level that encodes traffic rules and common sense. While some of these might be encoded using cost functions, such as maximizing distance from obstacles or ensuring smooth riding, other more complex behaviors, such as adapting driving in the presence of cyclists or properties of the ground, need to be implemented in an additional vertical layer that has access to all planning layers.

13.5   Coverage Path Planning

So far, we have only considered the problem of finding a (shortest) path. A variation of the path planning problem is *coverage path planning*. This is relevant for applications such as cleaning, mowing, or painting and usually aims at minimizing the time to completion and redundancy during coverage. This problem is closely related to the shortest path problem. For example, floor coverage can be achieved by performing a depth-first search (DFS) or a breadth-first search (BFS) on a graph where each vertex has the size of the coverage tool of the robot. "Coverage" is not only interesting for cleaning a floor: The same algorithms can be used to perform an exhaustive search of a configuration space, such as in the example shown in figure 3.3, where we plotted the error of a manipulator arm in reaching a desired position over its configuration space. Finding a minimum in this plot using an exhaustive search solves the inverse kinematics problem.

Doing a DFS or a BFS might generate efficient coverage paths, but they are far from optimal as many vertices might be visited twice. A path that connects all vertices in a graph but passes every vertex only once is known as a *Hamiltonian path*. A Hamiltonian path that returns to its starting vertex is known as a Hamiltonian cycle. This problem is also known as the traveling salesman problem (TSP), in which a route needs to be calculated that visits every city on the salesman's tour only once and is known to be nondeterministic polynomial (NP) time complete.

13.6   Summary and Outlook

Path planning is an ongoing research problem. Finding collision-free paths for mechanisms with high degrees of freedom (such as multiple arms operating in a shared space, multi-robot systems, or systems involving dynamics) is still a computationally intensive problem. Although sampling-based path planners can drastically speed up the time to find some solution, they are not optimal and struggle with algorithm-specific concerns such as navigating in narrow passages. There is no "silver bullet" algorithm for solving all path planning problems, and heuristics that lead to massive speed-up in one scenario might be detrimental in others. Also, algorithmic parameters are mostly ad hoc, and correctly tuning them to a specific environment may drastically increase performance.

Take-Home Lessons

• The first step in path planning is choosing a map representation that is appropriate to the application (chapter 12).

• The second step is to reduce the robot to a point-mass, which allows planning in the configuration space (or C-space).

• This allows the application of general-purpose shortest path graph-based algorithms, which have applications in a large variety of domains and are not limited to robotics.

• A sampling-based planning algorithm finds paths by sampling random points in the environment. Heuristics are used to maximize the exploration of space and bias the direction of search. This makes these algorithms fast but neither optimal nor complete.

• As the resulting paths are random, multiple trials might lead to entirely different results.

• There is no one-size-fits-all algorithm for path planning, and care must be taken to select the right paradigm (e.g., single-query versus multi-query), heuristics, and parameters.

Exercises

1. How does the computational complexity of Dijkstra's algorithm change when moving from 2D to 3D search spaces?

2. A* uses a "heuristic" to bias the search in the expected direction of the goal. Why can it only use a heuristic, not the actual length?

3. Assuming points are sampled uniformly at random in a randomized planning algorithm. Calculate the limiting behavior of the ratio (area of points in tree)/(area of points sampled) as the number of points sampled goes to infinity, assuming no duplicates. Assume the total area A_{total} and the area of free space A_{free} within are known.

4. Assuming a kd-tree is used as a nearest-neighbour data structure and that points are sampled uniformly at random, calculate the run-time of inserting a point into a tree of size N. Use "big-Oh" notation $\mathcal{O}(N)$.

5. What other practical run-time concerns must one consider besides computational complexity alone when doing sampling-based motion planning? Can you suggest ways to deal with these other concerns?

6. Write a program that can read in a simple map provided as a textfile where "1" indicates obstacles and "0" indicates free space.

a) Implement Dijkstra's algorithm to find the shortest path between any two given points in free space.

b) Implement A* to find the shortest path between any two given points.

c) How do these two implementations compare in terms of computational complexity?

7. Write a program that can read an image file in which white areas represent navigable space and black areas are obstacles. Implement a basic RRT algorithm to find the shortest path between any two points.

8. Explore the internet for libraries that implement "path planning" in the language of your choice. What tools do you find? How do they define the map? Do they perform obstacle avoidance? Does the kinematics of the robot matter?

9. Extend your path planning implementation for use with a differential-wheel robot. Describe steps that you would need to take for Dijkstra/A* and for RRT.

10. Extend your path planning algorithm for use with a two-link robot arm. Would you plan in joint or in configuration space, and what are advantages and drawbacks of each?

11. How does the computational complexity change when moving from a single six degrees of freedom (6-DoF) robot arm to a torso with two 6-DoF robots? Can you think about an approach that maintains the original computational complexity? What are the drawbacks of this approach.

12. Consider a robotic assembly task in which a robot retrieves objects from a known location and assembles them on a table. When can you rely on simple inverse kinematics and when do you need path planning?

13. How does a planning problem change when you not only consider positions but also forces and torques? Could you use a variation of RRT to solve such a problem?

14. Download a robotic path planning tool that allows you to try different algorithms.

a) Compare solution quality and speed among the different solutions.

b) What do you need to take into account when using randomized planners? Is comparing single-shot experiments sufficient?

15. Implement a coverage path planner for a single robot on a grid map using depth-first search. Evaluate the amount of redundancy for different starting locations.

14 Manipulation

While grasping (chapter 5) is generally concerned with connecting an object to a robot's kinematic chain, the act of grasping itself is usually only a small part of the tasks involved in physically dealing with objects.

> Think of all the possible ways in which you interact with objects on a daily basis. Identify which interactions can be classified as "grasping" and what is actually "manipulation"? How many times do you need to plan for a complex sequence of actions (e.g., making coffee in the morning)?

Oftentimes, the intention of the grasping action is to change the pose of this object in a precise, repeatable, and purposeful way. For example, cutlery and dishes need to be in well-specified areas and aligned with each other when setting a table; merchandise needs to be neatly stacked on a shelf; machine parts need to be assembled according to a specific order. These activities are more generally known as *manipulation*.

The goals of this chapter are to introduce the following concepts:

- The difference between grasping and manipulation
- Algorithms for choosing the right grasp
- Canonical manipulation tasks such as pick-and-place and assembly

14.1   Nonprehensile Manipulation

Manipulation can be thought of as a superset of grasping that includes additional capabilities that are typically referred to as *nonprehensile*—that is, anything but grasping. Indeed, objects can be pushed, poked, tossed, flipped, inserted, screwed-in, turned, twisted, and more. However, discussing all the possible ways objects might be manipulated and the many different contexts in which such actions would be required—which might dramatically change the approach a robot would need to choose—is well beyond the scope of this book and still a matter of active research.

Many manipulation problems can be cast into a sequence of pick-and-place problems in which the possible grasp choices are appropriately constrained. For example, an object can be turned or flipped by planning a sequence of pick-and-place movements that each turn the object by a certain degree. Similarly, using two robotic arms, with one grasping an object out of the hand of the other, will allow a robot system to change an object's pose almost arbitrarily. (Which poses an object will be able to reach will depend on the object's exact geometry, the kinematics of the robotic arms, and constraints in the workspace.) So-called *in-hand manipulation* is still an active area of research since repeatedly picking and placing an object and handovers between different arms is considered to be too slow and otherwise impractical for many application areas.

14.2   Choosing the Right Grasp

While we have so far only considered the mechanics of grasping (chapter 5), choosing an appropriate pose for grasping an object in a specific way is an algorithmic problem.

Finding a good grasp that fully constrains an object against all possible external forces and torques (i.e., a grasp that lies within the "grasping wrench space" detailed in section 5.1.1) may be too restrictive and, oftentimes, unnecessarily so. For example, it might be sufficient to find a grasp that constrains an object simply against gravity. Other applications might require the grasp to constrain an object's movement also against lateral forces that happen due to acceleration. In practice, these considerations usually lead to simple application-specific heuristics. For example, in warehouse picking tasks (Correll et al. 2016), the problem can be constrained to have the robot grasp only objects that are suitable to be retrieved with a simple suction cup. Finding a good grasp is then reduced to finding a flat surface close to the object's perceived center of gravity. When considering household tasks, such has handling and placing dishes, using silverware to pick up food, or holding a pitcher, we are often interested in very specific grasps that support the intended manipulation that follows.

Theoretically speaking, grasps such as picking up an object or opening a door by turning its knob are task-specific wrench spaces. We can then say that the grasp is "good" when the task wrench space is a subset of the grasping wrench space; otherwise it will fail. We can also look at the ratio between the forces actually applied to the object and the minimum force needed to perform a desired wrench. If this ratio is high—for example, when the robot grasps an object far from its center of gravity or has to squeeze an object heavily to prevent it from slipping—this grasp is not as good as one where the ratio is low, since in this latter case, all the force the robot is exerting is being efficiently utilized for the intended purposes. Unfortunately, it is usually not possible to find close-form expressions for the grasping wrench space. Rather, one can sample the space of suitable force vectors—for example, by picking a couple of forces that are on the boundary of the cone's base and calculating the convex hull over the resulting wrenches.

14.2.1 Finding Good Grasps for Simple Grippers

Finding good grasps for simple grippers—that is, those with only one or at most two degrees of freedom (DoFs) as covered in section 5.2—reduces the problem to finding geometries on the object that are suitable to place the gripper's jaws: That is, what we need is to find two parallel faces that are reasonably flat and at a distance that is lower than the gripper's maximum aperture. In practice, an object might be perceived by a three-dimensional (3D) perception device such as a stereo camera or a laser scanner, which provides only one perspective of an object and may introduce noise and uncertainty (chapter 15). A typical grasping pipeline using such a device is shown in figure 14.1 and proceeds as follows:

1. Acquisition: Obtain a "point cloud" or "depth image" of the objects of interest (figure 14.1b).

2. Preprocessing: Remove table plane or other points that are either too close or too far from the sensor (figure 14.1c).

3. Segmentation: Cluster points that are close enough to identify individual objects (figure 14.1d).

4. Filtering: Filter clusters by size, geometry, or other features, to down-select objects of interest (figure 14.1e).

5. Planning: Compute center-of-mass and principal axes of relevant clusters (figure 14.1f).

6. Collision checking: Generate possible grasps and check for collisions with point clouds (figure 14.1g).

7. Execution: Physically test a grasp by monitoring jaw distances as well as forces and torques at the wrist (figure 14.1h).

Some of these steps might not be necessary for all grasps, and some of them might become very complicated for some task. For example, preprocessing is often used to remove known quantities such as a table surface from the data, but it might become nontrivial when removing the edges of a bin or operating with a container of an arbitrary size.

Segmentation is the most critical step and requires some previous knowledge about the objects to grasp, such as their size or the geometry of features thereon. In figure 14.1, clustering points based on their distance is sufficient when, for example, using the DBSCAN algorithm (Ester et al. 1996), but it requires an assumption about object size in order to select a suitable threshold. Other segmentation algorithms might use surface normals, a combination of point cloud and image data such as color or patterns, or deep learning.

Filtering the resulting clusters to identify objects of interest can be as simple as rejecting those that are too small (as shown in figure 14.1e), but it might also involve matching

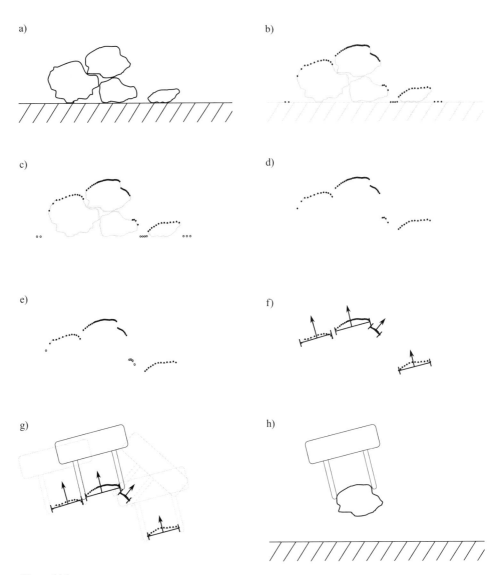

Figure 14.1
(a) Random objects on a table, (b) measurements from a laser scanner on the objects' surface, (c) removal of table plane, (d) connected components after segmentation, (e) removal of connected components based on size, (f) calculation of principal axes, (g) evaluation of possible grasps based on collisions, (h) physically attempting grasp.

the points to a 3D model of a desired object or making use of image data (e.g., ICP and RANSAC from chapter 12).

A simple approach to *plan* for possible grasps is to calculate the center-of-mass as well as the principal axes of an object using principal component analysis (appendix B.5). Other approaches might again require matching the existing points to a 3D model of the object to identify specific grasp points (such as the handle of a cup) or relying on image features to do so.

After planning all, or some, possible grasps, they need to be checked for *feasibility*. While a collision with a point in the point cloud might rule out a grasp, local search is sometimes being used to find a collision-free variant—for example, by (virtually) moving the gripper up and down as well as along the principal axes. In other applications—for example, bin picking—some collisions might be ignored with the expectation that the gripper will push other objects out of its way.

Even though a grasp might look robust in a point cloud representation, it might not be effective when physically executing it. Possible failures are collisions with objects, insufficient friction with the object, or an object moving before the gripper is fully closed. For this reason, it is important to close the gripper as much as possible before approaching the object, increasing the requirements for accurate perception.

With the recent advent of (deep) machine learning and the ability of neural networks (chapter 10) to approximate complex functions, it is also possible to replace parts or all of the algorithmic steps shown in figure 14.1 with a convolutional neural network trained by deep learning. While data intensive, such an approach can seamlessly merge image and depth data and adapt to application-specific data better than a hand-coded algorithm can.

14.2.2   Finding Good Grasps for Multifingered Hands

The simple grasping pipeline described above is computationally expensive because usually there are many possible grasp candidates and each of them needs to be checked for collisions. This problem becomes even more relevant when considering grippers with articulated fingers. This can be overcome by considering only a predefined set of grasps (e.g., two and three finger pinches for small objects and full-hand encompassing grasps for larger objects).

A suitable method to search the full space of possible grasps with an articulated hand is to use random sampling, such as moving the end-effector to random poses, closing its fingers around the object, and seeing what happens when generating wrenches that fulfill the task's requirements. "Seeing what happens" is usually performed first in simulation, and it requires collision checking and dynamic simulation. Dynamic simulation applies Newtonian mechanics to an object (i.e., forces lead to acceleration of a body) and moves the object at very small time-steps. While this can be done using the connected components identified in the point cloud alone and assuming reasonable parameters for friction and contact points, point cloud data can also be augmented by object models to simulate whether a grasp has a high likelihood to be successful. Here, there is a trade-off in exploring the space of possible grasps in simulation and actually trying grasps with the real hardware.

14.3   Pick and Place

One of the most basic manipulation problems is known as "pick and place," which involves grasping an object, transporting it, and placing it. However, what looks like a simple action is actually a sequence of individual tasks that can fail for multiple reasons. Pick and place consists of the following steps (figure 14.2):

1. Approaching the object
2. Grasping the object
3. Lifting the object
4. Moving the object to an intermediate pose
5. Placing the object
6. Releasing the object

   Each of these actions might not work as intended, requiring the robot to abort and restart the process. For example, what seems like a reliable grasp may turn out not to be suitable for actually lifting the object. Or a suitable path (chapter 13) toward the desired approach pose may not be found, making it be necessary to find another suitable approach pose first. This problem is known as *task and motion planning* (TAMP). Once a suitable approach pose has been found, placing the object will require the monitoring of forces and torques to ensure a gentle placement. Finally, releasing the object might require one to verify the intended pose.

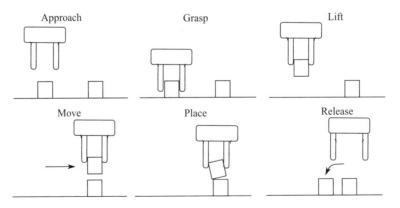

Figure 14.2
Various stages of pick-and-place (or grasping) from approach to release. Problems during an early step, such as during placing, as shown here, may lead to the failure of later stages of the algorithm.

As failure can occur anywhere in the above pipeline and manually encoding all possible state transitions will quickly get out of hand, behavior trees (section 11.4) have emerged as a powerful tool to encode complex sensor-based action sequences. A sample behavior tree for a picking task is shown in figure 11.7.

14.4   Peg-in-Hole Problems

A canonical manipulation problem and a special case of pick-and-place is the *peg-in-hole* problem and variations thereof—including hole-on-peg problems, which generalize insertion and assembly operations. Peg-in-hole requires repeatable grasping of an object and using force- and torque-based search motions to find the hole. Typically, insertion patterns consist of tilting motions and spiral-shaped search motions (Watson, Miller, and Correll 2020). Both approaches have advantages and drawbacks. For peg-in-hole tasks, tilt insertion tends to work better for objects larger than 1 centimeter (cm) in diameter.

Tilt insertion is detailed in figure 14.3 and proceeds as follows: (*a*) Given a hole pose, the part is held vertically above the hole by a preset distance; (*b*) the gripper is then tilted about its local *y*-axis and translated horizontally in the global frame in the direction corresponding to the hand's local *x*-axis; (*c*) this offset places the lowest part of the bottom edge of the part directly above the estimated center of the hole, and then the hand is translated downward in

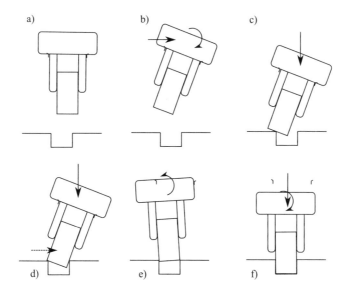

Figure 14.3
Tilt-based peg-in-hole insertion. Arrows indicate the directions of gripper translation and rotation. Dashed lines indicate motion that results from compliant grasping.

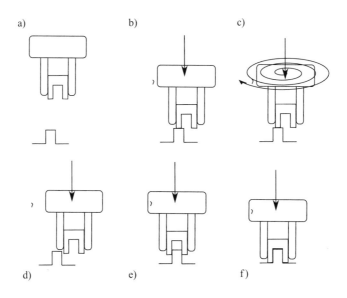

Figure 14.4
Spiral-based hole-on-peg assembly. Arrows indicate the directions of gripper translation and rotation. The spiral motion is performed in the horizontal plane and is accompanied by downward pressure. It either leads to missing the peg (*d*) or finding the peg (*e*) and complete assembly (*f*).

the global frame until the part makes contact with the hole; (*d*) naturally, the shaft cannot sink deeply into the hole when tilted this way but, rather, the curvature of the hole meets the circumferential surface of the shaft and gathers it toward the center of the hole, exploiting compliance in the gripper; (*e*) after being centered, the shaft is tilted back up a few degrees beyond vertical; (*f*) finally, the shaft is returned to true vertical and pushed down into the hole until a preset reaction force (commensurate with the hole fit) is met. If the z-component of the hand's final pose is near to the expected value, the insertion is considered successful.

Spiral insertion is best suited for hole-on-peg operations, and peg-in-hole operations with a peg diameter less than 1 cm. It is described for hole-on-peg assemblies in figure 14.4. The algorithm proceeds as follows: (*a*) Given a shaft end pose, the grasped part—for example, a gear with a center hole—is held vertically above the shaft by a preset distance. (*b*) The hand then moves down along the axis of insertion until a reaction force threshold is reached. (*c*) The robot then performs a spiraling motion defined in polar coordinates with its origin at the point of contact; the algorithm also probes the contact surface to determine whether the reaction force against the wrist in z-direction falls below a threshold, and it continues in this fashion until one of several conditions are met: (*d*) the gripper pose exceeds a threshold below which the intended hole pose cannot lie, or (*e*) lateral torques at the wrist exceed a threshold indicating the gear has slid onto the shaft and cannot translate laterally. The vertical probing steps also serve to sink the part into place before attempting a lateral move that might sabotage what would otherwise be a successful insertion. Once the torque threshold is met, the hand pushes down until a preset reaction force (commensurate with the hole's tolerance) is met. (*f*) If the *z*-component of the hand's final pose is near to the expected value, the assembly is considered a success.

Implementing peg-in-hole and hole-on-peg insertion also has a large number of possible failure modes, making a behavior tree (BT) architecture most suitable. However, force- and torque-based controls require operating at a higher bandwidth than a typical BT architecture would support—on the order of hundreds of hertz (Hz) and more. How to implement these closed-loop control loops is strongly dependent on the actual hardware. For example, some robot arms do not make raw force/torque values available at a rate that is required for real-time control but provide built-in functions to move until certain force limits are met. This allows the robot manufacturer to run real-time controllers that ensure safety, which would be more difficult to accomplish when providing low-level control to the user.

Take-Home Lessons

• Performing simple pick-and-place tasks is now possible thanks to high-resolution, in-hand sensing and fast-enough computation that can sift through large amounts of point cloud data in real time.

• Planning and executing a grasp is a complex problem that encompasses topics from many of the previous chapters, ranging from identifying an object, localizing it, and computing the inverse kinematics to reach it.

• Manipulation extends these techniques from the robot itself to the objects the robot deals with and how they relate to each other, and it remains an open field of research.

• Seemingly simple manipulation tasks such as pick-and-place or peg-in-hole assembly require both high-level planning and force-torque real-time control, posing a combined task and motion planning problem.

Exercises

1. Write code to generate rectangles with random dimensions and orientations. Rectangles can overlap. Use a point-in-polygon test to simulate random point samples on their surface, simulating a top-down view with a depth sensor.

a) Implement a segmentation routine that clusters objects based on a minimum distance.

b) Implement a filter that rejects connected components based on size. For which kind of objects does this work well, and where does this method fail?

c) Implement a filter that rejects connected components that do not have rectangular shape. Are you able to specify a filter that works independent of the object size?

d) Apply principal component analysis to compute the principal axes of the rectangle and compare with ground truth. How does the number of samples affect the accuracy of your estimate?

2. Use a function of the kind $u(x-i) + rand(j)$ with $u(x)$ the unit step function, $rand()$ uniformly distributed random noise, and i, j suitable parameters to simulate a noisy depth image of a cube with width i. Use the nearest neighbor of each point to compute its normals and a suitable clustering algorithm to identify the cube. How do i and j affect the accuracy of your estimate?

3. Think about simulating peg-in-hole assembly in a robotic simulator. What are the problems with using a simulation environment when simulating tight assembly problems?

4. Perform an internet search for "open source" robotic assembly problems and re-create them in your laboratory. Implement a spiral and tilt-based assembly controller.

IV UNCERTAINTY

15 Uncertainty and Error Propagation

Robots are systems that combine sensing, actuation, computation, and communication. All of its subsystems are subject to a high degree of uncertainty. This can be observed in daily life: Phone calls with poor signal make it hard to understand the other party, text characters are difficult to read from far away or at low resolution, the wheels of your car may slip when accelerating on a rainy road after a red light, or your neural network mistakes a cat for a dog. In robotics, measurements taken by onboard sensors are sensitive to changing environmental conditions and subject to electrical and mechanical limitations. Similarly, actuators are not accurate since joints and gears have backlash and wheels can slip. Finally, wireless communication in particular, whether using radio or infrared, is notoriously unreliable. Consider how these types of uncertainty are all different: Are they continuous or discrete? How does the uncertainty corrupt the "ideal"? How can these various types of uncertainties be quantified and accounted for? So far, we have considered uncertainty only as far as limitations in accuracy and precision, and we assumed that they do not matter. The goals of this chapter are to understand the following:

- How to treat uncertainty mathematically using probability theory
- How measurements with different uncertainty can be combined
- How error propagates when taking multiple measurements in a row

This discussion will help us to better understand how sensor error affects higher-level features and decisions, while also creating the basis for dealing with uncertainty and the problems that arise from it.

This chapter requires an understanding of random variables, probability density functions, and the normal (alternatively called the "Gaussian") distribution. These concepts are explained in a robotic sensing context in appendix C.1.

15.1 Uncertainty in Robotics as a Random Variable

As quantities such as "distance to a wall," "position on the plane," or "I can see a blue cross (yes/no)" are uncertain, we can consider them random variables. A *random variable* can be thought of as the outcome of a "random" experiment, such as the face shown when rolling a die or the speed of an individual molecule of gas in a room. Just because a variable is random does not mean we know nothing about it. For instance, we can roll a fair six-sided die hundreds of times and create a table of likelihoods of each side coming up. We can also measure the temperature in a room and understand the average speed of those gas molecules using the kinetic theory of gases.

Experiments in robotics rarely involve true statistical randomness because of their scale and design. Instead, there are two primary sources of uncertainty: sensors and physical interactions. Sensors are intrinsically noisy because of the physical phenomena associated with them. These sources of uncertainty are often modeled using Gaussian (or "normal") probability distribution functions, as they accurately model measurement uncertainty at large samples, per the central limit theorem. Moreover, Gaussian distributions are mathematically convenient for combining multiple noisy measurements and analyzing propagation of uncertainty. As individual sensor readings can be considered random variables, quantities derived from multiple sensors can be considered random variables as well. Also, some physical interactions are very challenging to model accurately, especially those involving friction, leading to uncertainty in the resulting models. This chapter focuses on how to characterize the uncertainty of such aggregated quantities from the uncertainty that characterizes the individual sensors and modeling assumptions.

15.2   Error Propagation

We'll begin with an example of error propagation that is a core motivation for needing to quantify uncertainty: the distance traveled by a differential-wheel robot given the rotations of its wheels. It turns out that the Gaussian distribution is very appropriate to model the uncertainty in this process. The robot moves with an expected displacement (e.g., as commanded by the motors on each wheel) plus some uncertain displacement that can be decomposed into the radial and tangential directions for each time-step as a result of wheel slip (see figure 15.1). We can say that *process noise* drawn from a Gaussian distribution was added to the position resulting from the commanded motion. This process noise has zero mean and distinct variance in both the radial and tangential directions; the process noise is a static property of the wheel-ground interaction.

Such a robot (when subject to slip) will actually increase the uncertainty in its position the farther it drives. Initially at a known location, the expected value (or mean) of its position will become increasingly uncertain, corresponding to an increasing position variance. This variance is obviously somehow related to the variance of the underlying mechanism (the process variance)—namely, the slipping wheel. Interestingly, we will see its position variance grow much faster orthogonal to the robot's direction of motion, as small errors in orientation have a much larger cumulative effect on position than small errors in the longitudinal direction, as illustrated in figure 15.1.

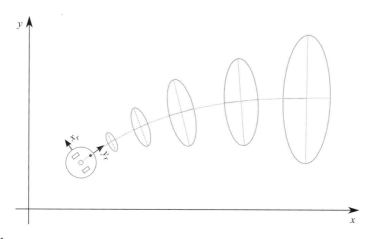

Figure 15.1
Two-dimensional normal distribution depicting growing uncertainty as the robot moves. Although starting with equal uncertainty in x and y, the large effect of small errors in orientation causes the error to grow faster in y-direction of the robot.

If only there were a way to correct this unbounded error with some sort of sensor measurement! However, even these sensor measurements would be affected by uncertainty, so we will have to take this into account as well. We'll come up with a correction that is able to accommodate this shortly.

Similarly, when estimating distance and angle to a wall (a line feature in two dimensions) from point cloud data, the uncertainty of the random variables describing distance and angle to the wall are related to the uncertainty of *each* point measured on the wall. These relationships are formally captured by the *error propagation law*.

The key intuition behind the error propagation law is that the variance of each component that contributes noise to a random variable has a weight associated with it. This weight is a function of how strongly that component influences the random variable. Measurements that have little effect on the aggregated random variable should also have little effect on its variance, and vice versa. "How strongly" one variable affects another can be expressed by the ratio of how small changes of the first variable relate to small changes in the second one. This concept should sound familiar, as it is none other than the partial derivative of the first variable with respect to the second. For example, let $y = f(x)$ be a function that maps a random variable x (a sensor reading) to a random variable y (a feature). Let the standard deviation of x be given by σ_x. We can then calculate the resulting standard deviation σ_y by

$$\sigma_y = \left(\frac{\partial f}{\partial x} \right) \sigma_x \tag{15.1}$$

and its variance σ_y^2 as

$$\sigma_y^2 = \left(\frac{\partial f}{\partial x}\right)^2 \sigma_x^2. \tag{15.2}$$

In case $y = f(\mathbf{x})$ is a multivariable function that maps n inputs to m outputs, variances between these variables may be represented by *covariance matrices*, a representation of the various combinations in which these variables may affect one another, one by one. A covariance matrix holds the variance of each n input variable along its diagonal and is zero otherwise, if the random variables are not correlated. We can then write

$$\Sigma^Y = J \Sigma^X J^T, \tag{15.3}$$

where Σ^X and Σ^Y are the $n \times n$ and $m \times m$ covariance matrices holding the variances of the input and output variables, respectively, and J is a $m \times n$ *Jacobian* matrix, which holds the partial derivatives $\frac{\partial f_i}{\partial x_j}$. As J has n columns, each row contains partial derivatives with respect to x_1 to x_n.

15.2.1 Example: Line Fitting

Detecting walls is a fairly common sensing task for a mobile robot, and it can be accomplished with a range-bearing sensor like a lidar. We can make some simplifying assumptions for the sake of this next example—namely, that a wall sensed by a two-dimensional (2D) spinning lidar would appear as a straight line of points in three-dimensional (3D) space. Thus, our wall detection problem can be described as estimating the angle α and distance r of a line (the wall) from a set of points (lidar readings) given by (ρ_i, θ_i) using equations (9.4) and (9.5). We can now express the relationship of changes of ρ_i to changes in α by

$$\frac{\partial \alpha}{\partial \rho_i}. \tag{15.4}$$

Similarly, we can calculate $\frac{\partial \alpha}{\partial \theta_i}$, $\frac{\partial r}{\partial \rho_i}$, and $\frac{\partial r}{\partial \theta_i}$. We can do this because we have derived analytical expressions for α and r as a function of θ_i and ρ_i in chapter 9.

We are now interested in deriving equations for calculating the variance of α and r as a function of the variances of the distance measurements. Let's assume each distance measurement ρ_i has variance $\sigma_{\rho,i}^2$ and each angular measurement θ_i has variance $\sigma_{\theta,i}^2$. We now want to calculate $\sigma_{\alpha,i}^2$ as the weighted sum of $\sigma_{\rho,i}^2$ and $\sigma_{\theta,i}^2$, each weighted by its influence on α:

$$\sigma_{\alpha,i}^2 = \frac{\partial \alpha_i^2}{\partial \rho_i} \sigma_{\rho,i}^2 + \frac{\partial \alpha_i^2}{\partial \theta_i} \sigma_{\theta,i}^2. \tag{15.5}$$

Derivation of $\sigma_{r,i}^2$ is done similarly.

More generally, if we have I input variables X_i and K output variables Y_k, the covariance matrix of the output variables σ_Y can be expressed as $\sigma_Y^2 = \frac{\partial f^2}{\partial X} \sigma_X^2$, where σ_X is the covariance matrix of input variables and J is a Jacobian matrix of a function f that calculates Y

from X and has the form

$$J = \frac{\partial \mathbf{f}}{\partial \mathbf{X}} = \begin{bmatrix} \frac{\partial f_1}{\partial X_1} & \cdots & \frac{\partial f_1}{\partial X_I} \\ \vdots & \ddots & \vdots \\ \frac{\partial f_K}{\partial X_1} & \cdots & \frac{\partial f_K}{\partial X_I} \end{bmatrix}. \tag{15.6}$$

15.2.2 Example: Odometry

Whereas the line-fitting example demonstrated a many-to-one mapping (where multiple instantaneous measurements form a feature), odometry requires calculating the variance that results from multiple sequential measurements. Error propagation allows us to not only express the robot's position but also the variance of this estimate. For this task, our list of questions to answer is as follows:

1. What are the input variables and what are the output variables?
2. What are the functions that calculate output given an input?
3. What is the variance of the input variables?

As usual, we describe the robot's position by a tuple (x, y, θ). These are the three output variables. We can measure the distance each wheel travels Δs_r and Δs_l based on the encoder ticks and the known wheel radius. These are the two input variables. We can now calculate the change in the robot's position by calculating

$$\Delta s = \frac{\Delta s_r + \Delta s_l}{2} \tag{15.7}$$

$$\Delta x = \Delta s \, \cos(\theta) \tag{15.8}$$

$$\Delta y = \Delta s \, \sin(\theta) \tag{15.9}$$

$$\Delta \theta = \frac{\Delta s_r - \Delta s_l}{b}. \tag{15.10}$$

The robot's new position is then given by

$$f(x, y, \theta, \Delta s_r, \Delta s_l) = [x, y, \theta]^T + [\Delta x, \Delta y, \Delta \theta]^T. \tag{15.11}$$

We now have a function that relates our measurements to our output variables. What makes things complicated here is that the output variables are also a function of their previous values. Therefore, their variance does not only depend on the variance of the input variables but also on the previous variance of the output variables. We therefore need to write

$$\Sigma_{p'} = \nabla_p f \Sigma_p \nabla_p f^T + \nabla_{\Delta_{r,l}} f \Sigma_\Delta \nabla_{\Delta_{r,l}} f^T. \tag{15.12}$$

The first term is the error propagation from an initial position $p = [x, y, \theta]$ to a new position p'. For this we need to calculate the partial derivatives of f with respect to x, y, and θ. This is a 3×3 matrix:

$$\nabla_p f = \begin{bmatrix} \frac{\partial f}{\partial x} & \frac{\partial f}{\partial y} & \frac{\partial f}{\partial \theta} \end{bmatrix} = \begin{bmatrix} 1 & 0 & -\Delta s \sin(\theta + \Delta\theta/2) \\ 0 & 1 & \Delta s \cos(\theta + \Delta\theta/2) \\ 0 & 0 & 1 \end{bmatrix}. \quad (15.13)$$

The second term is the error propagation of the actual wheel slip. This requires calculating the partial derivatives of f with respect to Δs_r and Δs_l, which is a 3×2 matrix. The first column contains the partial derivatives of x, y, θ with respect to Δs_r. The second column contains the partial derivatives of x, y, θ with respect to Δs_l:

$$\nabla_{\Delta_{r,l}} f = \begin{bmatrix} \frac{1}{2}\cos\left(\theta + \frac{\Delta\theta/2}{b}\right) - \frac{\Delta s}{2b}\sin\left(\theta + \frac{\Delta\theta}{b}\right) & \frac{1}{2}\cos\left(\theta + \frac{\Delta\theta/2}{b}\right) - \frac{\Delta s}{2b}\sin\left(\theta + \frac{\Delta\theta}{b}\right) \\ \frac{1}{2}\sin\left(\theta + \frac{\Delta\theta/2}{b}\right) + \frac{\Delta s}{2b}\cos\left(\theta + \frac{\Delta\theta}{b}\right) & \frac{1}{2}\sin\left(\theta + \frac{\Delta\theta/2}{b}\right) + \frac{\Delta s}{2b}\cos\left(\theta + \frac{\Delta\theta}{b}\right) \\ \frac{1}{2} & -\frac{1}{2} \end{bmatrix}.$$

$$(15.14)$$

Finally, we need to define the covariance matrix for the measurement noise. As the error is proportional to the distance traveled, we can define Σ_Δ by

$$\Sigma_\Delta = \begin{bmatrix} k_r|\Delta s_r| & 0 \\ 0 & k_l|\Delta s_l| \end{bmatrix}. \quad (15.15)$$

Here, k_r and k_l are constants that need to be found experimentally and $|\cdot|$ indicates the absolute value of the distance traveled. We also assume that the error of the two wheels is independent, which is expressed by the zeros in the matrix.

We now have all ingredients for equation 15.12, allowing us to calculate the covariance matrix of the robot's pose just as in figure 15.1.

15.3   Optimal Sensor Fusion

We have now seen how errors from *different* sources can propagate into error of compound measurements by means of the equations that relate input to output error. We are now interested in how independent observations of the *same* quantity can be combined. For example, we have considered measurements obtained from two different wheels that are combined in a pose estimate. What about a case in which we receive two independent measurements of the robot's pose? Similarly, we have seen how to combine multiple point measurements into a line. How about two observations of the same line (distance and angle) from two different sensors?

Let $\hat{q}_1$ and $\hat{q}_2$ be two different estimates of a random variable and σ_1^2 and σ_2^2 their variances, respectively. Let q be the true value. This could represent the true pose of a line, with observations having different variances when they are obtained by different means—say, using a lidar for $\hat{q}_1$ and by using a camera for $\hat{q}_2$. We can now define the weighted mean-square error:

$$S = \sum_{i=1}^{n} \frac{1}{\sigma_i} (q - \hat{q}_i)^2. \tag{15.16}$$

That is, S is the sum of the errors of each observation $\hat{q}_i$ with $n = 2$, weighted by the inverse of their standard deviation $\frac{1}{\sigma_i}$.

Each error is weighted with the inverse of its standard deviation to put more emphasis on observations whose uncertainty is low. Minimizing S by taking the derivatives of S with respect to $\hat{q}_i$ and setting them to zero yields the following optimal expression for q:

$$q = \frac{\hat{q}_1 \sigma_2^2}{\sigma_1^2 + \sigma_2^2} + \frac{\hat{q}_2 \sigma_1^2}{\sigma_1^2 + \sigma_2^2} \tag{15.17}$$

or, equivalently,

$$q = \hat{q}_1 + \frac{\sigma_1^2}{\sigma_1^2 + \sigma_2^2} (\hat{q}_2 - \hat{q}_1). \tag{15.18}$$

We have now derived an expression for fusing two independent observations with different variances that provably minimizes the error between our estimate and the real value. As q is a linear combination of two random variables (section C.4), the new variance is given by

$$\sigma^2 = \frac{1}{\frac{1}{\sigma_1^2} + \frac{1}{\sigma_2^2}}. \tag{15.19}$$

Interestingly, the resulting variance is smaller than both σ_1 and σ_2; that is, incorporating additional observations can always help improve accuracy instead of introducing more uncertainty.

15.3.1   The Kalman Filter

Although we have introduced the problem above as fusing two observations of the same quantity and weighting them by their variance, we can also interpret the equation above as an update step that calculates a new estimate of an observation based on its old estimate and a new measurement. Specifically, we can interpret the expression $\hat{q}_2 - \hat{q}_1$ from equation (15.18) as the difference between what the robot actually sees and what it thinks it should see. This term is known as *innovation* in what is also known as the *Kalman filter*. We can now rewrite (15.18) from above into

$$\hat{x}_{k+1} = \hat{x}_k + K_{k+1} \tilde{y}_{k+1}, \tag{15.20}$$

also known as the *perception update step*. Here, $\hat{x}_k$ is the state we are interested in at time k, $K_{k+1} = \frac{\sigma_1^2}{\sigma_1^2 + \sigma_2^2}$ is what is known as the *Kalman gain*, and $\tilde{y}_{k+1} = \hat{q}_2 - \hat{q}_1$ is the innovation.

Unfortunately, there are few systems that allow us to directly measure the information we are interested in. Rather, we obtain a sensor measurement z_k that we need to convert into something we can use to update our state. We can then consider the inverse problem of predicting your measurement z_k from your state x_k. This is done using the observation model H_k, so that

$$\tilde{y}_k = z_k - H_k x_k, \tag{15.21}$$

where $H_k x_k$ is the measurement prediction. In our example, H_k was just the identity matrix; in a robot position estimation problem, H_k is a function that would predict how a robot would observe a shift in position through a sensor. As you can see, all the weighting based on variances is done in the Kalman gain K.

It is now time for a brief disclaimer: The Kalman filter only works for linear systems. Forward kinematics of even the simplest robots are mostly intrinsically nonlinear, and so are observation models that relate sensor observations to the robot position. Nonlinear systems can be dealt with by using the *extended Kalman filter*, which will be introduced a bit later on in the context of robot localization.

Take-Home Lessons

- Uncertainty can be expressed by means of a probability density function.
- More often than not, the Gaussian distribution is chosen because it allows treating error with powerful analytical tools.
- In order to calculate the uncertainty of a variable that is derived from a series of measurements, we need to calculate a weighted sum in which each measurement's variance is weighted by its impact on the output variable. This impact is expressed by the partial derivative of the function relating input to output.
- It is also possible to fuse independent observations, each with their own variance, of the same quantity. This will usually reduce the variance of the resulting observation.

Exercises

1. Given two observations $\hat{q}_1$ and $\hat{q}_2$ with variances σ_1 and σ_2 of a normal distributed process with actual value $\hat{q}$, an optimal estimate can be calculated by minimizing the expression

$$S = \frac{1}{\sigma_1^2}(\hat{q} - \hat{q}_1)^2 + \frac{1}{\sigma_2^2}(\hat{q} - \hat{q}_2)^2.$$

Calculate $\hat{q}$ so that S is minimized.

2. An ultrasound sensor measures distance $x = c\Delta t/2$. Here, c is the speed of sound and Δt is the difference in time between emitting and receiving a signal.

a) Let the variance of your time measurement Δt be σ_t^2. What can you say about the variance of x, when c is assumed to be constant? (Hint: How does a change in Δt affect x?)

b) Now assume that c is changing depending on location and weather, as examples, and can be estimated with variance σ_c^2. What is the variance of x now?

3. Consider a unicycle that turns with angular velocity $\dot{\phi}$ and has radius r. Its speed is thus a function of $\dot{\phi}$ and r and is given by

$$v = f(\dot{\phi}, r) = r\dot{\phi}.$$

Assume that your measurement of $\dot{\phi}$ is noisy and has a standard deviation $\sigma_{\dot{\phi}}$. Use the error propagation law to calculate the resulting variance of your speed estimate σ_v^2.

4. Consider a scenario in which a robot can localize itself against landmarks. Describe what happens to the robot's positional error in the following three cases:

a) The landmark location is known and the robot can reliably localize to it.

b) The landmark location has a variance and the robot can reliably localize to it.

c) Both the landmark and localization mechanism have a variance.

5. Write a program that graphically illustrates merging observations with two different variances in one dimension and two dimensions (1D and 2D).

16 Localization

Robots employ sensors and actuators that are subject to uncertainty. Chapter 15 describes how to quantify this uncertainty using probability density functions that associate a probability with each possible outcome of a random process, such as the reading of a sensor or the actual physical change of an actuator. Here, the robot's pose is a compound metric that is of central importance to mobile robotics, and that is the focus of this chapter.

There are many ways to localize a robot in its environment, and odometry is just one of them. A different possible way to localize a robot in its environment is to extract high-level features (chapter 9), such as the distance to a wall, from a number of different sensors.

As we have seen in chapter 15, uncertainty keeps propagating without the ability to take corrective measurements. The goals of this chapter are to present mathematical tools and algorithms that will allow you to actually shrink the uncertainty of a measurement by combining it with additional observations. In particular, this chapter covers the following topics:

- Using landmarks to improve the accuracy of a discrete position estimate (Markov localization and Bayes filter)
- Approximating continuous position estimates (particle filter)
- Using the extended Kalman filter (EKF) to estimate a continuous position estimate

16.1 Motivating Example

Imagine a floor with three doors, two of which are closer together and the third farther down the corridor (figure 16.1). Imagine now that your robot is able to detect doors—namely, that it is able to tell whether it is in front of a wall or in front of a door. Features like this can serve as a landmark for the robot. Given a map of this simple environment and no information whatsoever about where our robot is located, we can use landmarks to drastically reduce the space of possible locations once the robot has passed one of the doors. One way of representing this belief is to describe the robot's position with three Gaussian distributions, each centered in front of a door and its variance a function of the uncertainty with which

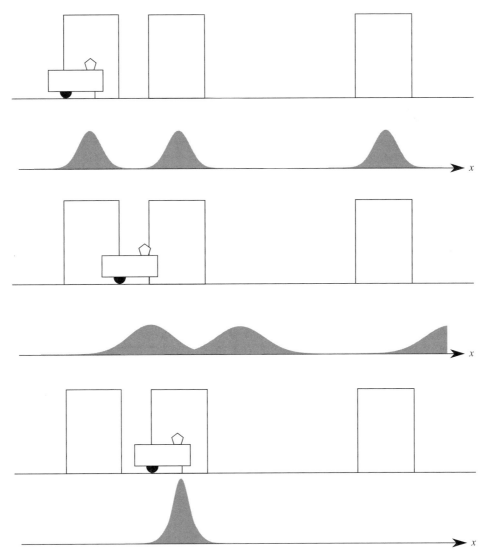

Figure 16.1
A robot localizing itself using a "door detector" in a known map. Top: Upon encountering a door, the robot can be in front of any of the three doors. Middle: When driving to the right, the Gaussian distributions representing its location also shift to the right and widen, representing growing uncertainty. Bottom: After detecting the second door, the robot can discard hypotheses that are not in front of the door and gain certainty on its location.

the robot can detect a door's center. This is known as a multi-hypothesis belief, since we have a hypothesis stating that the robot can be in front of each door. What happens if the robot continues to move? From the error propagation law we know the following:

1. The Gaussians describing the robot's three possible locations will move with the robot.
2. The variance of each Gaussian will keep increasing with the distance the robot moves.

What happens if the robot arrives at another door? Given a map of the environment, we can now map the three Gaussian distributions to the location of the three doors. Because all three Gaussians will have moved but the doors are not equally spaced, only some of the peaks will coincide with the location of a door. Assuming we trust our door detector much more than our odometry estimate, we can now remove all beliefs that do not coincide with a door. Again assuming our door detector can detect the center of a door with some accuracy, our location estimate's uncertainty is now only limited by that of the door detector.

Things are just slightly more complicated if our door detector is also subject to uncertainty: There is a chance that we are in front of a door but haven't noticed it. Then, it would be a mistake to remove this belief. Instead, we just weigh all beliefs with the probability that there *could* be a door. Say that our door detector detects false-positives with a 10 percent chance. Then, there is a 10 percent chance of being at any location that is not in front of a door, even if our detector tells us we are in front of a door. Similarly, our detector might detect false-negatives with 20 percent chance, telling us that there is no door, even though the robot is just in front of it. Thus, we would need to weigh all locations in front of a door with 20 percent chance and all locations not in front of a door with 80 percent likelihood if our robot tells us there is no door, even if we are indeed in front of one.

16.2 Markov Localization

Calculating the probability of being at a certain location given the likelihood of certain observations is the same as any other conditional probability. There is a formal way to describe such situations: Bayes' rule (appendix C.2):

$$P(A|B) = \frac{P(A)P(B|A)}{P(B)}. \tag{16.1}$$

16.2.1 Perception Update

How does this map into a localization framework? Let's assume event A is equivalent to being at a specific location *loc*. Let's also assume that event B corresponds to the event of seeing a particular feature *feat*. We can now rewrite Bayes' rule to be

$$P(loc|feat) = \frac{P(loc)P(feat|loc)}{P(feat)}. \tag{16.2}$$

Rephrasing Bayes' rule in this way, we can calculate the probability of being at location *loc*, given that we see feature *feat*. This is known as *perception update*. For example, *loc* could correspond to door 1, 2, or 3, and *feat* could be the event of sensing a door. What do we need to know to make use of this equation?

1. We need to know the prior probability of being at location loc $P(loc)$.

2. We need to know the probability of seeing the feature if we were actually at this location $P(feat|loc)$.

3. We need the probability of encountering the feature feat $P(feat)$.

Let's start with the third requirement, which might be the most confusing part of information we need to collect. It may make more sense to consider $P(feat) = \sum_{x \in locations} P(feat|x) * P(x)$, the probability that we'd see this feature in a given location for every possible location. It is also common to see this term set to 1, with $P(loc|feat)$ written as being proportional to the numerator of equation 16.2 instead of equals.

The prior probability of being at location *loc*, $P(loc)$, is called the *belief model*. In the case of the three-door example, it is the value of the Gaussian distribution underneath the door corresponding to *loc*.

Finally, we need to know the probability $P(feat|loc)$ of seeing the feature *feat*, given that we are at location *loc*. If your sensor was perfect, this probability is simply 1, if the feature exists at this location, or 0, if the feature cannot be observed at this location. If your sensor is not perfect, $P(feat|loc)$ corresponds to the likelihood of the sensor to see the feature if it exists.

The last missing piece involves deciding how to represent possible locations. In the graphical example in figure 16.1, we assumed Gaussian distributions for each possible location. Alternatively, we can discretize the world into a grid and calculate the likelihood of the robot being in any of its cells. In our three-door world, it might make sense to choose grid cells that have the width of a door.

16.2.2   Action Update

One of the assumptions in the above thought experiment was that we know with certainty that the robot moved right. We will now more formally study how to treat uncertainty from motion. Recall that odometry input is just another sensor that we assume to have a Gaussian distribution; if our odometer tells us that the robot traveled a meter, it could have traveled a little less or a little more, with decreasing likelihood the further we get from the given measurement. We can therefore calculate the posterior probability of the robot moving from a position loc' to loc, given its odometer input *odo*:

$$P(loc' -> loc|odo) = P(loc' -> loc)P(odo|loc' -> loc)/P(odo). \qquad (16.3)$$

This is again Bayes' rule. The unconditional probability $P(loc' -> loc)$ is the prior probability for the robot to have been at location loc'. The term $P(odo|loc' -> loc)$ corresponds to the probability of getting an odometer reading *odo* after traveling from a position *loc′* to *loc*. If getting a reading of the amount *odo* is reasonable for the distance from *loc′* to *loc*, this probability is high. If it is unreasonable—for example, if the distance is larger than what is physically possible—this probability should be very low.

As the robot's location is uncertain, the real challenge is now that the robot could have potentially been anywhere to start with. We therefore have to calculate the posterior probability $P(loc|odo)$ for all possible positions *loc′*. This can be accomplished by summing over all possible locations:

$$P(loc|odo) = \sum_{loc'} P(loc' -> loc)P(odo|loc' -> loc). \tag{16.4}$$

In other words, the law of total probability requires us to consider all possible locations the robot could have ever been at. This step is known as the *action update*. In practice, we don't need to calculate this for all possible locations but only those that are technically feasible, given the maximum speed of the robot. We note also that the sum notation technically corresponds to a convolution (appendix C.3) of the probability distribution of the robot's location in the environment with the robot's odometry error probability distribution.

16.2.3 Example: Markov Localization on a Topological Map

We have now learned two methods to update the belief distribution of where the robot could be in the environment. First, a robot can use external landmarks to update its position. This is known as the *perception update* and relies on exteroception. Second, a robot can observe its internal sensors. This is an instance of an *action update* and relies on proprioception. The combination of action and perception updates is known as *Markov localization*. You can think about the action update as increasing the uncertainty of the robot's position and the perception update as shrinking it. (You can also think about the action update as a discrete version of the error propagation model.)

To illustrate this, we now describe one of the first successful real robot systems that employed Markov localization in an office environment. The experiment is described in more detail in a 1995 article of *AI Magazine* (Nourbakhsh, Powers, and Birchfield 1995). The office environment consisted of two rooms and a corridor that can be modeled by a topological map (figure 16.2). In a topological map, areas that the robot can be in are modeled as vertices, and navigable connections between them are modeled as edges of a graph. The location of the robot can now be represented as a probability distribution over the vertices of this graph.

The robot has the following sensing abilities:

- It can detect a closed door to its left or right.
- It can detect an open door to its left or right.
- It can detect whether it is an open hallway.

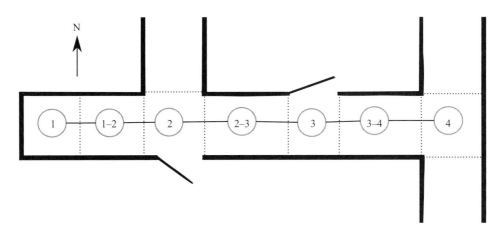

Figure 16.2
An office environment consisting of two rooms connected by a hallway. A topological map is superimposed.

Table 16.1
Conditional probabilities of the Dervish robot detecting certain features in the Stanford laboratory.

	Wall	Closed dr	Open dr	Open hwy	Foyer
Nothing detected	70%	40%	5%	0.1%	30%
Closed door detected	30%	60%	0%	0%	5%
Open door detected	0%	0%	90%	10%	15%
Open hallway detected	0%	0%	0.1%	90%	50%

Unfortunately, the robot's sensors are not at all reliable. The researchers have experimentally found the probabilities to obtain a certain sensor response for specific physical positions using their robot in their environment. These values are provided in table 16.1.

For example, the success rate to detect a closed door is only 60 percent, whereas a foyer looks like an open door in 15 percent of the trials. This data corresponds to the conditional probability to detect a certain feature given a certain location.

Consider now the following initial belief state distribution: $p('1\text{--}2') = 0.8$ and $p('2\text{--}3') = 0.2$. Here, '1–2' and '2–3' refer to the positions on the topological map in figure 16.2. For this domain, we are told with certainty that the robot faces east. The robot now drives for a while until it reports "open hallway on its left and open door on its right." This actually corresponds to location 2, but the robot can in fact be anywhere. For example, there is a 10 percent chance that the open door is in fact an open hallway (i.e., the robot is really at position 4). How can we calculate the new probability distribution of the robot's location? Here are the possible trajectories that could happen.

The robot could move from 2–3 to 3, 3–4, and finally 4. We have chosen this sequence as the probability to detect an open door on its right is zero for 3 and 3–4, which leaves position 4 as the only option if the robot has started at 2–3. In order for this hypothesis to be true, the following events need to have happened, with their probabilities given in parentheses:

1. The robot must have started at 2–3 (20 percent).

2. The robot must not have seen the open door at the left of 3 (5 percent) and not have seen the wall at the right (70 percent).

3. The robot must not have seen the wall to its left (70 percent) and not have seen the wall to its right at node 3–4 (70 percent).

4. The robot must correctly identify the open hallway to its left (90 percent) and mistake the open hallway to its right for an open door (10 percent).

Together, the likelihood that the robot got from position 2–3 to position 4 is therefore given by $0.2 \times 0.05 \times 0.7 \times 0.7 \times 0.7 \times 0.9 \times 0.1 = 0.03\%$, which is very unlikely.

The robot could also move from 1–2 to 2, 2–3, 3, 3–4, or 4. We can evaluate these hypotheses in a similar way:

• The chance that it correctly detects the open hallway and door at position 2 is 0.9×0.9, so the chance of being at position 2, having started at 1–2, is $0.8 \times 0.9 \times 0.9 = 64\%$.

• The robot cannot have ended up at position 2–3, 3, and 3–4 because the chance of seeing an open door instead of a wall on the right side is zero in all these cases.

• In order to reach position 4, the robot having started at 1–2 has a chance of 0.8. The robot must not have seen the hallway on its left and the open door to its right when passing position 2. The probability for this is 0.001×0.05. The robot must then have detected nothing at 2–3 (0.7×0.7), nothing at 3 (0.05×0.7), nothing at 3–4 (0.7×0.7), and finally mistaken the hallway on its right for an open door at position 4 (0.9×0.1). Multiplied together, this outcome is very unlikely.

Given this information, we can now calculate the posterior probability of being at a certain location on the topological map by adding up the probabilities for every possible path to get there.

16.3 The Bayes Filter

We have seen how sensor measurements can be formally incorporated into a position estimate using Bayes' rule, which relates the likelihood of being at a certain position given that the robot sees a certain feature to the likelihood of the robot seeing this feature if it were really at the hypothetical location. We have also seen how the robot can use its sensor model to relate its observation with possible positions. Its real location is likely to be somewhere between its original belief (based on error propagation) and where the sensor tells it that it is. We will now provide an algorithm for localizing a robot through a multi-hypothesis, iterative process that does not depend on a particular class of motion or sensor model (e.g., the Gaussian noise models used by Kalman filters).

To formalize our terms and notation, we will describe our robot's motion model as the distribution given by $P(x'|x, u)$—that is, the probability of being in a particular state x' given that we started in state x and executed action u. We can describe our sensor model as being characterized by the distribution given by $P(z|x)$—namely, the probability that we would see sensor observation z if we were in state x. This is not limited to discrete locations, as in the previous sensor, but could also be the likelihood of an ultrasound sensor detecting a wall at certain distance. Typically, this will require some discretization of the environment, such as a grid. Finally, we will define the probability of being in a particular state x as $P(x)$.

Our goal with the Bayes filter will be to estimate our robot's state over time (x_t, where t indicates time-step) given a history of actions and observations (sensor measurements). To do so, we will compute the posterior probability of our state estimate, also known as *belief*, using this history. We define the belief that our robot is in state x at time t given a history of actions ($u_1, \ldots u_t$) and sensor measurements ($z_1, \ldots, z_t$) as

$$Bel(x_t) = P(x_t|u_1, z_1, u_2, z_2, \ldots, u_t, z_t).$$

By leveraging the Markov assumption, that our current state only depends on our previous state x_{t-1} and action u_t, we can greatly simplify the computation required as such:

$$P(x_t|x_{0:t-1}, z_{1:t-1}, u_{1:t}) = P(x_t|x_{t-1}, u_t).$$

For example, if we wanted to calculate the probability of an observation z_t, we know that the only term that actually matters is the robot's current state (since the other terms don't affect what sensor readings we'd expect to get):

$$P(z_t|x_{0:t}, z_{1:t-1}, u_{1:t}) = P(z_t|x_t).$$

We will now derive a recursive definition for belief that makes iteratively computing state belief over a time history of actions and observations tractable. Beginning with our initial definition of belief, we will apply Bayes' rule, the Markov property, the law of total probability, and recursion to achieve our goal. We will use c to denote the normalizing constant (from the denominator of Bayes' rule), which is the same for all possible x_t, as follows:

$$Bel(x_t) = P(x_t|u_1, z_1, \ldots, u_t, z_t) \tag{16.5}$$

$$Bel(x_t) = c * P(z_t|x_t, u_1, z_1, \ldots, u_t, z_t) * P(x_t|u_1, z_1, \ldots, u_t) \tag{16.6}$$

$$Bel(x_t) = c * P(z_t|x_t) * P(x_t|u_1, z_1, \ldots, u_t) \tag{16.7}$$

$$Bel(x_t) = c * P(z_t|x_t) *$$
$$\sum_{x_{t-1} \in X} P(x_t|u_1, z_1, \ldots, u_t, x_{t-1}) * P(x_{t-1}|u_1, z_1, \ldots, z_{t-1}, u_t) \tag{16.8}$$

$$Bel(x_t) = c * P(z_t|x_t) * \sum_{x_{t-1} \in X} P(x_t|u_t, x_{t-1}) * P(x_{t-1}|u_1, z_1, \ldots, z_{t-1}) \qquad (16.9)$$

$$Bel(x_t) = c * P(z_t|x_t) * \sum_{x_{t-1} \in X} P(x_t|u_t, x_{t-1}) * Bel(x_{t-1}) \qquad (16.10)$$

This final equation is remarkable because it allows us to perform a belief update for a given state by incorporating a sensor measurement and/or a motion prediction based on an action we took. With this formulation, we can define an algorithm for belief updates that takes our current belief, an array of action and observation data, and the set of states that comprise the state space as inputs, returning an updated belief that incorporates this information:

```
BayesFilter(Belief Bel, Data d, Set of States X):
  while d is not empty:
    c = 0
    if (d[0] is a sensor measurement):
      z = d.pop(0)
      for all x ∈ X:
        Bel'(x) = P(z|x)Bel(x)
        c += Bel'(x)
      for all x ∈ X:
        Bel'(x) = c⁻¹*Bel'(x)
    elif (d[0] is an action):
      u = d.pop(0)
      for all x ∈ X:
        Bel'(x) = Σₓₜ₋₁ P(x|u,xₜ₋₁)*Bel(xₜ₋₁)
    Bel = Bel'
  return Bel
```

This powerful idea of iteratively incorporating sensor measurements and motion predictions underpins an entire family of state estimation methods. In the sections that follow, we will extend this concept to be applicable in contexts that we often find robots: infinitely large, continuous state spaces that we cannot exhaustively iterate over.

16.3.1 Example: Bayes Filter on a Grid

Instead of using a coarse topological map, we can also model the environment as a fine-grained grid. Each cell is marked with a probability corresponding to the likelihood of the robot being at this exact location (figure 16.3). We assume that the robot is able to detect walls with some certainty, perhaps with a short-range ultrasonic sensor on the front, back, and sides of the robot. The images in the right column show the actual location of the robot, while the left column shows the probability of the robot being in each grid cell. Initially, the robot does not see a wall and therefore could be almost anywhere. The robot now moves northward. The action update now propagates the probability of the robot being

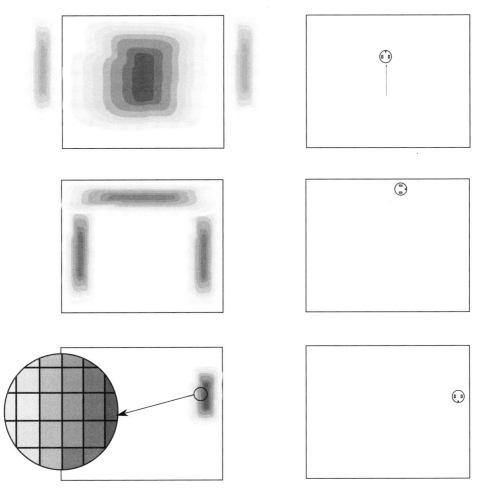

Figure 16.3
Markov localization on a grid. The left column shows the likelihood of being in a specific cell as a gray value (dark colors correspond to high likelihoods). The right column shows the actual robot location. Arrows indicate previous motion. Initially, the position of the robot is unknown, but recorded upward motion makes positions at the top of the map more likely. After the robot has encountered a wall, positions away from walls become unlikely. After rightward and down motions, the possible positions have shrunk to a small area.

somewhere north. As soon as the robot encounters the wall, the perception update bumps up the likelihood of being higher in grid cells near walls. Because there is some uncertainty associated with the wall detector, the robot will not only have likelihood directly at the wall but also at other locations—with decreasing probability—close by to walls. As the action update involved continuous motion to the north, the likelihood that the robot is close to the south wall is almost zero. The robot then performs a right turn and travels along the wall in the clockwise direction. As soon as it hits the east wall, it is almost certain about its position, which then again decreases as the robot continues to travel.

16.4   Particle Filter

Although grid-based Markov localization can provide compelling results, it can be computationally very expensive, in particular when the environment is large and the resolution of the grid is small. This is in part because we need to carry the probability of being at a certain location forward for every cell on the grid, regardless of how small this probability is. An elegant solution to this problem is the particle filter. It works as follows:

1. Represent the robot's position by N particles that are randomly distributed around its estimated initial position. For this, we can either use one or more Gaussian distributions around the initial estimate(s) of where the robot is or choose a uniform distribution (figure 16.4).

2. Every time the robot moves, we will move each particle in the exact same way but add noise to each movement, much like we would observe on the real robot. Without a perception update, the particles will spread apart farther and farther.

3. Upon a perception event, we evaluate every single particle using our sensor model. What is the likelihood of having a perception event such as we observed at this location? We can then use Bayes' rule to update each particle's position.

4. Once in a while or during perception events that render certain particles infeasible, particles that have a probability that is too low can be deleted, while those with the highest probability can be replicated.

Step 1. Observation   Let us now assume that we can detect line features $z_{k,i} = (\alpha_i, r_i)^T$, where α and r are the angle and distance of the line from the coordinate system of the robot. These line features are subject to variances $\sigma_{\alpha,i}$ and $\sigma_{r,i}$, which make up the diagonal of R_k. See the line detection section for a derivation of how angle and distance as well as their variance can be calculated from a laser scanner. The observation is a 2×1 matrix.

Step 2. Measurement update   We assume that we can uniquely identify the lines we are seeing and retrieve their real position from a map that we have been given in advance. This is much easier for unique features but can also be done for lines by assuming that our error is small enough and we therefore can search through our map and pick the closest lines. As features are stored in global coordinates, we need to transform them into how the

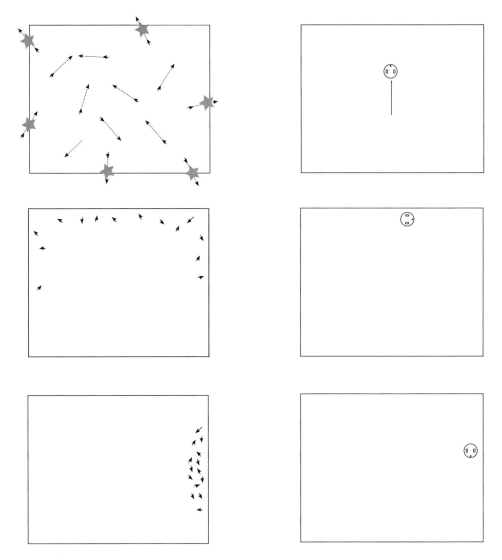

Figure 16.4
Particle filter example. Possible positions and orientations of the robot are initially uniformly distributed. Particles move based on the robot's motion model. Particles that would require the robot to move through a wall in the absence of a wall perception event are deleted (stars). After a perception event, particles too far from a wall become too unlikely and are resampled to be in the vicinity of a wall. Eventually, the particle filter converges.

robot would see them. In practice, this is nothing but a list of lines, each with an angle and a distance, but this time with respect to the origin of the global coordinate system. Transforming them into robot coordinates is straightforward. With $\hat{x}_k = (x_k, y_k, \theta_k)^T$ and $m_i = (\alpha_i, r_i)$ the corresponding entry from the map, we can write

$$h(\hat{x}_{k|k-1}) = \begin{bmatrix} \alpha_{k,i} \\ r_{k,i} \end{bmatrix} = h(x, m_i) = \begin{bmatrix} \alpha_i - \theta \\ r_i - (x\cos(\alpha_i) + y\sin(\alpha_i)) \end{bmatrix} \tag{16.11}$$

and calculate its Jacobian $\boldsymbol{H}_k$ as the partial derivatives of α to x, y, θ in the first row, and the partial derivatives of r in the second. How to calculate $h()$ to predict the radius at which the robot should see the feature with radius r_i from the map is illustrated in the figure below.

Step 3. Matching   We are now equipped with a measurement z_k and a prediction $h(\hat{x}_{k|k-1})$ based on all features stored in our map. We can now calculate the innovation

$$\tilde{y}_k = z_k - h(\hat{x}_{k|k-1}), \tag{16.12}$$

which is simply the difference between each feature that we can see and those that we predict from the map. The innovation is again a 2×1 matrix.

A major strength of particle filters is that they are nonparametric estimators of arbitrary probability distributions and thus are able to accommodate nonlinear functions that Kalman filters cannot. However, this is not the only algorithm available for utilizing nonlinear motion and sensor models for state estimation. We now introduce a modification to the Kalman filter enabling the use of nonlinear models.

16.5   Extended Kalman Filter

In contrast to the linear models required of the Kalman filter, in the extended Kalman filter the state transition and observation models do not need to be linear functions of the state but may instead be any function as long as it's differentiable. The action prediction step looks as follows:

$$\hat{x}_{k'|k-1} = f(\hat{x}_{k-1}, \boldsymbol{u}_{k-1}). \tag{16.13}$$

Here, $f()$ is a function of the previous state x_{k-1} and control input $\boldsymbol{u}_{k-1}$. A good example for such an equation is the odometry update we are already familiar with. Here, $f()$ is a function describing the forward kinematics of the robot, with x_k its position and $\boldsymbol{u}_k$ the wheel speed we set.

Sticking with our well-known example, we can also calculate the covariance matrix of the robot position as follows:

$$\boldsymbol{P}_{k'|k-1} = \nabla_{x,y,\theta} f \boldsymbol{P}_{k-1|k-1} \nabla_{x,y,\theta} f^T + \nabla_{\Delta_{r,l}} f \boldsymbol{Q}_{k-1} \nabla_{\Delta_{r,l}} f^T, \tag{16.14}$$

where $\boldsymbol{Q}_k$ was the covariance matrix of the wheel-slip and the Jacobian matrices of the forward kinematic equations $f()$ with respect to the robot's position (indicated by the index x, y, θ) and with respect to the wheel-slip of the left and right wheel.

The perception update step now looks as follows:

$$\hat{x}_{k|k'} = \hat{x}_{k'|k-1} + \boldsymbol{K}_{k'}\tilde{\boldsymbol{y}}_{k'} \tag{16.15}$$

$$\boldsymbol{P}_{k|k'} = (I - \boldsymbol{K}_{k'}\boldsymbol{H}_{k'})\boldsymbol{P}_{k'|k-1}. \tag{16.16}$$

We are calculating everything twice: Once we update from $k-1$ to an intermediate result k' during the action update using our motion model, we obtain the final result after performing the perception update where we go from k' to k.

We need to calculate three additional variables:

1. The innovation $\tilde{\boldsymbol{y}}_k = \boldsymbol{z}_k - h(\hat{\boldsymbol{x}}_{k|k-1})$
2. The covariance of the innovation $\boldsymbol{S}_k = \boldsymbol{H}_k \boldsymbol{P}_{k|k-1} \boldsymbol{H}_k^\top + \boldsymbol{R}_k$
3. The (near-optimal) Kalman gain $\boldsymbol{K}_k = \boldsymbol{P}_{k|k-1} \boldsymbol{H}_k^\top \boldsymbol{S}_k^{-1}$

Here, $h()$ is the observation model and $\boldsymbol{H}$ its Jacobian. How these equations are derived is involved (and is one of the fundamental results in control theory), but the idea is the same as introduced above: we wish to minimize the error of the prediction.

16.5.1   Odometry Using the Kalman Filter

We will show how a mobile robot equipped with a laser scanner that has a map of the environment can correct its position estimate by relying on unreliable odometry and unreliable sensing, in an optimal way. Whereas the update step is equivalent to forward kinematics and error propagation that we have seen before, the observation model and the calculation of the "innovation" require additional steps to perform odometry.

Step 1. Prediction   We assume for now that the reader is familiar with calculating $\hat{x}_{k'|k-1} = f(x, y, \theta)^T$ and its variance $\boldsymbol{P}_{k'|k-1}$. Here, $\boldsymbol{Q}_{k-1}$, the covariance matrix of the wheel-slip error, is given by

$$\boldsymbol{Q}_{k-1} = \begin{bmatrix} k_r|\Delta s_r| & 0 \\ 0 & k_l|\Delta s_l| \end{bmatrix}, \tag{16.17}$$

where Δs_l and Δs_r is the wheel movement of the left and right wheel and k_l and k_r are constants. Refer to the odometry lab for detailed derivations of these calculations and how to estimate k_r and k_l. The state vector $\hat{x}_{k'|k-1}$ is a 3×1 vector, the covariance matrix $\boldsymbol{P}_{k'|k-1}$ is a 3×3 matrix, and $\nabla_{\Delta_{r,l}}$ that is used during error propagation is a 3×2 matrix. See the error propagation section for details on how to calculate $\nabla_{\Delta_{r,l}}$.

Step 2. Observation   Let us now assume that we can detect line features $z_{k,i} = (\alpha_i, r_i)^T$, where α and r are the angle and distance of the line from the coordinate system of the robot. These line features are subject to variances $\sigma_{\alpha,i}$ and $\sigma_{r,i}$, which make up the diagonal of R_k. See the line detection section for a derivation of how angle and distance as well as their variance can be calculated from a laser scanner. The observation is a 2×1 matrix (representing angle and distance).

Step 3. Measurement update We assume that we can uniquely identify the lines we are seeing and retrieve their real position from a map. This is much easier for unique features but can also be done for lines by assuming that our error is small enough and that we can search through our map and pick the closest lines. As features are stored in global coordinates, we need to transform them into how the robot would see them. In practice, this is nothing but a list of lines specified with respect to the origin of the global coordinate system, each with an angle and a distance. Transforming them into robot coordinates is straightforward, following the development of (16.11).

Step 4. Matching We are now equipped with a measurement z_k and a prediction $h(\hat{x}_{k|k-1})$ based on all features stored in our map. We can now calculate the innovation

$$\tilde{y}_k = z_k - h(\hat{x}_{k|k-1}), \tag{16.18}$$

which is simply the difference between each feature that we can actually see (our sensor measurement) and the measurement values that we would expect if making a prediction using the map (not using our sensors). The innovation is again a 2×1 matrix.

5. Estimation We now have all the ingredients to perform the perception update step of the Kalman filter:

$$\hat{x}_{k|k'} = \hat{x}_{k'|k-1} + K_{k'}\tilde{y}_{k'} \tag{16.19}$$

$$P_{k|k'} = (I - K_{k'}H_{k'})P_{k'|k-1}. \tag{16.20}$$

It will provide us with an update of our position that fuses our odometry input and the information that we can extract from features in the environment in a way that takes into account their variances. That is, if the variance of your previous position is high (because you have no idea where you are) but the variance of your measurement is low (maybe from a GPS or a highly recognizable symbol on the wall), the Kalman filter will put more emphasis on your sensor. If your sensors are poor (maybe because you cannot tell different lines/walls apart), more emphasis will be placed on the odometry.

As the state vector is a 3×1 vector and the innovation a 2×1 matrix, the Kalman gain must be a 3×2 matrix. This can also be seen when looking at the covariance matrix that must come out as a 3×3 matrix and knowing that the Jacobian of the observation function is a 2×3 matrix. We can now calculate the covariance of the innovation and the Kalman gain using

$$S_k = H_k P_{k|k-1} H_k^\top + R_k \tag{16.21}$$

$$K_k = P_{k|k-1} H_k^\top S_k^{-1}. \tag{16.22}$$

16.6   Summary: Probabilistic Map-Based Localization

In order to localize a robot using a map, we need to perform the following steps:

1. Calculate an estimate of our new position using the forward kinematics and knowledge of the wheel speeds that we sent to the robot until the robot encounters some uniquely identifiable feature.

2. Calculate the relative position of the feature (e.g., a wall, a landmark or beacon) to the robot.

3. Use knowledge of where the feature is located in global coordinates to predict what the robot should see.

4. Calculate the difference between what the robot actually sees and what it believes it should see (e.g., using a Kalman filter).

5. Use the result from step 4 to update its belief by weighing each observation against its variance.

Steps 1 and 2 are based on the sections on "forward kinematics" (section 3.1) and "line detection" (section 9.3). Step 3 uses again simple forward kinematics to calculate the position of a feature stored in global coordinates in a map in robot coordinates. Step 4 is a simple subtraction of what the sensor sees and what the map says. Step 5 may induce the Kalman filter or an error minimization constraint.

Take-Home Lessons

• If the robot has no additional sensors and its odometry is noisy, error propagation will lead to ever-increasing uncertainty of a robot's position regardless of using Markov localization or the Kalman filter.

• Once the robot is able to sense features with known locations, Bayes' rule can be used to update the posterior probability of a possible position. The key insight is that the conditional probability of being at a certain position given a certain observation can be inferred from the likelihood of actually making this observation given a certain position.

• A complete solution that performs this process for discrete locations is known as Markov localization.

• The extended Kalman filter is the optimal way to fuse observations of different random variables that are Gaussian distributed.

• Possible random variables could be the estimate of your robot position from odometry and observations of static beacons with known location (but uncertain sensing) in the environment.

• In order to take advantage of the approach, you will need differentiable functions that relate measurements to state variables as well as an estimate of the covariance matrix of your sensors.

• An approximation that combines benefits of Markov localization (multiple hypothesis) and the Kalman filter (continuous representation of position estimates) is the particle filter.

Exercises

1. Assume that the ceiling is equipped with infrared markers that the robot can identify with some certainty. Your task is to develop a probabilistic localization scheme, and you would like to calculate the probability $p(marker|reading)$ of being close to a certain marker given a certain sensing reading and information about how the robot has moved.

a) Derive an expression for $p(marker|reading)$ assuming that you have an estimate of the probability to correctly identify a marker $p(reading|marker)$ and the probability $p(marker)$ of being underneath a specific marker.

b) Now assume that the likelihood that you are reading a marker correctly is 90 percent, that you get a wrong reading is 10 percent, and that you do not see a marker when passing right underneath it is 50 percent. Consider a narrow corridor that is equipped with four markers. You know with certainty that you started from the entry closest to marker 1 and move right in a straight line. The first reading you get is "marker 3." Calculate the probability that you are indeed underneath marker 3.

c) Could the robot also possibly be underneath marker 4?

17 Simultaneous Localization and Mapping

Robots are able to keep track of their position and orientation, known as *pose*, using a model of the noise arising in their drivetrain and their forward kinematics to propagate this error into a spatial probability density function (section 15.2). If the robot sees uniquely identifiable landmarks with known locations, the variance of this distribution would shrink. This can be accomplished for discrete locations using Bayes' rule (section 16.2) and for continuous distributions using the extended Kalman filter (section 16.5). The key insight here is that every observation will reduce the variance of the robot's position estimate. The Kalman filter performs an optimal fusion of two observations by weighting them inversely by their variance (i.e., unreliable observations count less than reliable ones). In the robot localization problem, one of the observations typically comes from the robot's proprioceptive position observations (e.g., using wheel encoders or feed-forward control inputs) whereas the other observation comes from a landmark with known location on a map composed of "landmarks." So far, we have assumed that these locations are known. This chapter explains the following:

- The concept of covariance (or what all the non-diagonal elements in the covariance matrix describe)
- How to estimate the robot's location and that of landmarks in the map at the same time (simultaneous localization and mapping, or SLAM)

17.1 Introduction

The SLAM problem has been a cornerstone problem of autonomous mobile robotics for a long time. It provides the foundation for a robot to be transported to an unknown location so that it can explore the area and build metrically accurate maps and pose estimates through onboard sensing alone. This could be useful for any field robot, whether terrestrial, extraterrestrial, under the ocean, or in an unexplored built environment. This chapter introduces one of the first comprehensive solutions to the problem to build understanding, even though it has since been superseded by computationally more efficient versions with a variety of algorithmic speed-ups and accuracy improvements. Let's begin by studying a series of special cases.

17.1.1   Landmarks

Since exteroceptive sensor measurements occur at a high frequency and generally must be processed in some way to reduce this data to algorithmically usable content, these measurements are generally distilled into features (chapter 9). The features present in each sensor measurement, as explained in section 9.3 for line recognition, are less numerous than the number of data points in each measurement and may be matched across measurements repeatably despite slight viewpoint changes. Worth noting here is that not all features may be matched across a sequence of sensor measurements. Those features that may be matched reliably represent coherent structures in the world (e.g., a wall or edge) and are geometrically pertinent. The structures are known as *landmarks*; they are geometric objects in the real world that can be used to inform motion within the world.

17.1.2   Special Case I: One Landmark

Consider an environment that you know has only a single landmark, but the position of the landmark is unknown. We assume that the robot is able to obtain the relative range and angle of this landmark, each with some variance. This landmark could be a tower but also a graphical tag that the robot can uniquely identify. The position of this measurement of the landmark $m_i = [\alpha_i, r_i]$ in global coordinates is unknown but can be calculated if an estimate of the robot's position $\hat{x}_k$ is known. The variance of m_i's components is now the variance of the robot's position plus the variance of the observation.

Now consider the robot moving toward the landmark and obtaining additional observations of it. Although the robot's position uncertainty is growing as it moves, it can now rely on the landmark m_i to reduce the variance of its prior position (as long as the landmark is stationary). Repeated observations of the same landmark from different angles and distances might improve the quality of its estimation of the landmark's position and hence its own position. The robot therefore has a chance to keep its variance very close to that with which it initially observed the landmark and stored it into its map!

How does the arithmetic operation of this measurement fusion proceed? As you may have already guessed, this probabilistic updating can be accomplished through the use of the extended Kalman filter (EKF) framework from section 16.5. In that treatment, we assumed that landmarks have a deterministic location but that the robot's sensing introduces a variance. This variance was propagated into the covariance matrix of the innovation (S). We can now simply add the variance of the estimate of the landmark's position to that of the robot's sensing process.

17.1.3   Special Case II: Two Landmarks

Consider now a map that has two landmarks. Visiting one after the other, the robot will be able to store both of them in its map, although the second landmark's location will be observed with higher variance because of the increasing positional variance with time. Although the observations of both landmarks are independent from each other, the relationship between their variances depend on the trajectory of the robot. The differences between these two variances are much lower if the robot observes them by moving in a straight line than when it performs a series of turns between them, since turns introduce greater variance in position.

As a thought experiment, consider this: a robot is driving for quite a long time and has accumulated a large variance in its position. It then observes the landmarks, one after the other, in a short period of time. The result of this would be that the probability density function over the distance between the two landmarks would have to be narrowly distributed. This probability density function can be understood as the *covariance* of the two random variables (each consisting of range and angle). In probability theory, the covariance is the measure of how much two variables are changing with respect to one another. Obviously, the covariance between the locations of two landmarks that are visited immediately after each other by a robot is much larger in magnitude than if those landmarks were to be observed far apart. This does not indicate that there is greater uncertainty in the landmarks' locations; rather, it means that there is correlation between the variables. It should therefore be possible to use the covariance between landmarks to correct estimates of landmarks in retrospect. If the robot returns to the first landmark it has observed, it will be able to reduce the variance of its position estimate. As it knows that it has not traveled very far since it observed the last landmark, it can then correct this landmark's position estimate.

17.2   The Covariance Matrix

When estimating quantities with multiple variables, such as the position of a robot that consists of its x-position, its y-position, and its orientation, matrix notation is a convenient way of writing down the relationships between them. For error propagation, we have written the variances of each input variable into the diagonal of a covariance matrix. For example, when using a differential-wheel robot, uncertainty in position expressed by σ_x, σ_y, and σ_θ were grounded in the uncertainty of its left and right wheel. We entered the variances of the left and right wheel into a 2×2 matrix and obtained a 3×3 matrix that had σ_x, σ_y, and σ_θ on its diagonal. Here, we set all other entries of the matrix to zero and ignored entries in the resulting matrix that were not on its diagonal. The reason we could do this is because the uncertainties in the left and right wheels are independent random processes: There is no reason that the left wheel slips just because the right wheel slips. Thus the covariance—the measure on how much two random variables are changing together—is zero. This is not the case for the robot's position: Uncertainty in one wheel will affect all output random variables (σ_x, σ_y, and σ_θ) at the same time, which is expressed by their nonzero covariances. Therefore, there will be nonzero entries off the diagonal of the output covariance matrix.

In the context of SLAM, we will maintain the poses of all landmarks that the robot is aware of in a column vector. There variances will make up the diagonal of a large covariance matrix. As the robot visits consecutive landmarks their variances are correlated, leading to nonzero diagonal entries.

17.3   EKF SLAM

The key idea in EKF SLAM is to extend the state vector from the robot's position (and potentially pose) to contain the position of all landmarks. Thus, the state

$$\hat{x}_{k'|k-1} = (x, y, \theta)^T, \tag{17.1}$$

becomes

$$\hat{x}_k = (x, y, \theta, \alpha_1, r_1, \ldots, \alpha_N, r_N)^T, \tag{17.2}$$

assuming N landmarks, which is a $(3 + 2N) \times 1$ vector. The action update (or "prediction update") is the same as if the landmarks are already known; the robot simply updates its position using odometry and updates the variance of its position using error propagation. The covariance matrix is now a $(3 + 2N) \times (3 + 2N)$ matrix that initially holds the variances on position and those of each landmark on its diagonal.

What about the perception update? Here it is worth noting that only one landmark is observed at a time; even if they are observed at nearly identical times, the algorithm requires only observing one landmark first and then the next. Thus, if the robot observes multiple landmarks at once, one needs to do multiple, consecutive perception updates. In practice, this implies that only those values of the observation vector—a $(3 + 2N) \times 1$ vector—that correspond to the landmark that you observe will be nonzero. Similar considerations apply to the observation function and its Jacobian.

17.3.1   Algorithm

We now introduce the algorithm for EKF SLAM, which is based on an iterative re-approximation scheme of the state vector and its corresponding covariance matrix. The state vector now includes the robot's position (potentially pose) and the position of all landmarks. The process proceeds as follows.

Initialization
To initialize the state vector, first set all of its entries to zeros. Begin by assuming no landmarks in the environment using the equation

$$x_0 = (0, 0, 0)^T \tag{17.3}$$

and set its covariance to a small number ϵ:

$$P_0 = \begin{bmatrix} \epsilon & 0 & 0 \\ 0 & \epsilon & 0 \\ 0 & 0 & \epsilon \end{bmatrix}. \tag{17.4}$$

The reason is that one cannot know any quantity definitively and also that otherwise the zero matrix would be uninvertible.

Update

If the robot is under motion and its sensors are providing information on landmarks of the map, then the state vector will be augmented in time while the pose of the robot will also be updated for each time-step. In EKF SLAM, both the sensor model and the process model may be nonlinear, so we will need to calculate the Jacobian of these functions with respect to the state and covariance matrices.

This update is a two-step process: first the prediction update, followed by the perception update.

Prediction update. If f is the nonlinear transition model for the system and u are hypothetical control inputs, then the state prediction update is given as

$$\hat{x}_{k'|k-1} = f(\hat{x}_{k-1}, u_{k-1}). \tag{17.5}$$

Meanwhile, the covariance prediction update is

$$\hat{P}_{k'|k-1} = F_{\hat{x}_{k-1}} P_{k-1} F_{\hat{x}_{k-1}}^T + N, \tag{17.6}$$

where $F_{\hat{x}_{k-1}} = \frac{\partial f(x,u)}{\partial x}|_{k-1}$ denotes the Jacobian matrix of the nonlinear transition model with respect to the state variable evaluated at $k-1$, and N is the covariance matrix of noise affecting the system's actuators (assumed to be additive in the state).

A remarkable result is that only the robot's state (and *not* the landmark positions in the world frame) is dependent on k, so most of $\hat{P}_{k'|k-1}$ will not be updated in this step.

Perception update. We assume that the sensor observation function $h(x)$ may be non-linear and is affected by additive noise with covariance R. The actual noisy measurement coming from the sensors will be denoted as y_k. The Jacobian of the observation function evaluated at the prior time-step is $H_{\hat{x}_{k'}} = \frac{\partial h(x)}{\partial x}|_{k'}$. This update operates on the results from the prediction update in order to provide a "fully fused" state estimate. The state perception update, which results in the next state estimate and corresponding state covariance estimate, is

$$\hat{x}_{k|k-1} = \hat{x}_{k'|k-1} + K_{k'}(y_k - h(\hat{x}_{k'|k-1})) \tag{17.7}$$

and

$$\hat{P}_{k|k-1} = \hat{P}_{k'|k-1} - K_{k'} Z_{k'} K_{k'}^T, \tag{17.8}$$

where $\boldsymbol{Z}_{k'}$ and $\boldsymbol{K}_{k'}$ are, respectively,

$$\boldsymbol{Z}_{k'} = \boldsymbol{H}_{\hat{\boldsymbol{x}}_{k'}} \hat{\boldsymbol{P}}_{k'|k-1} \boldsymbol{H}_{\hat{\boldsymbol{x}}_{k'}}^{T} + \boldsymbol{R} \tag{17.9}$$

and

$$\boldsymbol{K}_{k'} = \hat{\boldsymbol{P}}_{k'|k-1} \boldsymbol{H}_{\hat{\boldsymbol{x}}_{k'}}^{T} \boldsymbol{Z}_{k'}. \tag{17.10}$$

Crucially, this step requires the careful and error-free association of features with landmarks in a process known as *data association*. Data association can be accomplished in one of two ways: by using description vectors over the features and matching if the similarity of the descriptors are above a certain threshold, or through use of a Hungarian algorithm for optimal assignment. This data association step can be avoided if the landmarks are uniquely labeled and do not rely on feature matching. It may also depend on the generation of new landmarks or the deletion of old ones to keep the state vector of bounded length.

Note that there are some significant computational speed-ups associated with the matrix inversions and multiplications in the preceding update equations because of their sparsity, which we do not cover explicitly here. In total, the total complexity for the updates is $\mathcal{O}(kn^2)$, where k is the number of landmarks and n is the number of states. There are ways to in fact make this algorithm constant-time through a process known as *marginalization* (Sibley, Matthies, and Sukhatme 2010), wherein some variables are effectively no longer re-estimated (e.g., old poses that are no longer affecting the current pose significantly).

17.3.2   Multiple Sensors

The use of sensors in robotics is riddled with multiple cost-benefit analyses. A vision sensor provides information on the structure of the environment and pose-to-pose information (e.g., through frame-to-frame alignment); however, it succumbs in low-light or texture-less environments where frame alignment may no longer be feasible because of inadequate information content. A depth sensor can provide information on structure but can fail in many odometric tasks because of redundant geometry (e.g., in a building corridor). Finally, an (IMU can provide short-term odometry estimates intrinsically, but without any environmental sensing, quickly drifts and diverges. There are constantly more sensors being brought into the mix of this cost-benefit analysis, but one can only add so many sensors to a platform because of size, weight, power, and cost constraints. The choice of sensors must be made based on the expected environment and design considerations.

Once a sub-selection of sensors has been determined, these sensors can be integrated into an EKF SLAM system through a simple augmentation. In this case, the update step in section 17.3.1 may be augmented with an arbitrary number of sensors in order to be robust to failure modes in any one sensor. For instance, an IMU can provide orientation observations interleaved with a range sensor providing orientation and position observations based on landmarks. The overall algorithm for EKF SLAM is unchanged here, but merely introduces more update steps at different frequencies.

17.4   Graph-Based SLAM

Usually, a robot obtains an initial estimate of where it is using some onboard sensors (e.g., odometry, optical flow), leverages this estimate to localize landmarks (e.g., walls, corners, graphical patterns) in the environment, and finally refines its pose estimate by matching sensor information in consecutive fields-of-view using, for example, the iterative closest point (ICP) algorithm (section 12.2) or feature matching. As soon as a robot revisits the same landmark twice, it can update the estimate on its location. As consecutive observations are not independent but rather closely correlated, the refined estimate can then be propagated along the robot's path. This is formalized in EKF-based SLAM, where new measurements that are reliable can correct for errors in prior measurements.

A more intuitive way to look at this is to consider it as a "graph" made up of masses at nodes and springs on the edges. Consider this spring-mass analogy: Each possible pose (mass) is constrained to its neighboring pose by a spring. The higher the uncertainty of the relative transformation between two poses (e.g., obtained using odometry), the weaker the spring. Every time a robot gains confidence on a relative pose, the spring is stiffened instead. Eventually, all poses will be pulled in place so as to minimize the overall tension across the graph. This can be achieved by numerically minimizing the overall error based on all available observations using gradient descent. This formulation of the SLAM problem is known as *graph-based SLAM* (Grisetti et al. 2010). An example pose graph with feature-based landmarks is shown in figure 17.1.

17.4.1   SLAM as a Maximum-Likelihood Estimation Problem

The classical formulation of SLAM describes the problem as maximizing the posterior probability of all points on the robot's trajectory given the odometry input and the observations. Formally,

$$p(x_{1:T}, m | z_{1:T}, u_{1:T}), \tag{17.11}$$

where $x_{1:T}$ represents all discrete positions from time $t \in (1, T)$, z the observations, and u the odometry measurements. This formulation makes heavy use of the temporal structure of the problem. In practice, solving the SLAM problem has two requirements:

1. A motion update model—which is the probability $p(x_t | x_{t-1}, u_t)$ of being at location x_t given an odometry measurement u_t and being at location x_{t-1}

2. A sensor model—which is the probability $p(z_t | x_t, m_t)$ to make observation z_t given the robot is at location x_t and the map m_t

Note that these are reminiscent of the probabilities we invoked on action and perception updates in the EKF. Namely, in EKF SLAM, we maintained a probability density function for the robot pose as well as the positions of all landmarks on the map. Being able to address the data association problem, where features correspond with landmarks, is still of utmost

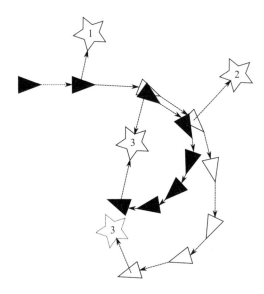

Figure 17.1
Robot poses (triangles) and unique landmarks (stars) form a pose graph on a two-dimensional (2D) map. Edges between robot poses indicate odometry measurements. Edges between robot poses and landmarks indicate range and bearing measurements. Upon a loop-closure, here the rediscovery of landmark 3, all poses between the events can be adjusted.

importance. Like EKF-based SLAM, graph-based SLAM does not solve this problem and will fail if landmarks are confused.

In graph-based SLAM, a robot's trajectory forms the nodes of a graph whose edges are transformations (translation and rotation) that have a variance associated with them. An alternative view is the spring-mass analogy mentioned above. Instead of having each spring wiggle a node into place, graph-based SLAM aims at finding those locations that maximize the joint likelihood of all observations. Said differently, of all the possible values the state variables can take, graph-based SLAM finds the "best" values that are defined as the *most likely ones* based on the evidence (observations from odometry and sensors). As such, graph-based SLAM is a *maximum-likelihood estimation* problem.

To ground this arithmetically, let's revisit the normal distribution:

$$\frac{1}{\sigma\sqrt{2\pi}} e^{\frac{-(x-\mu)^2}{2\sigma^2}} . \tag{17.12}$$

It provides the probability for a measurement to have value x given that this measurement is normal distributed with mean μ and variance σ^2. This is a univariate (single-variable) formulation of the normal distribution; however, it can be extended to multivariate distributions by considering x and μ as vectors and σ as the covariance matrix. We can now associate such a distribution with every node-to-node transformation. We denote the measurement of a transformation between node i and a node j as z_{ij}. Its expected value is denoted as $\hat{z}_{ij}$;

this expected value is based on some model for the measurements that operate on the state variables and output an "expected measurement" based on the current value of the state variables.

A quick aside regarding expected measurements: We've seen this before. Recall that in section 8.4 we constructed a sensor model for a camera of 3D points based on 2D measurements. Those 3D points here represent landmarks we are tracking, and the 2D measurements are the $\hat{z}_{ij}$ of constraints between the position and the projection of the landmarks. Note that the projection of the landmarks as a sensor measurement are 2D in the camera case, so there is a correspondence between the real measurement taken of a landmark z_{ij} and the predicted one $\hat{z}_{ij}$ based on the estimated position of the robot and the estimated position of the landmark.

Formulating a normal distribution of measurements z_{ij} with mean $\hat{z}_{ij}$ and a covariance matrix Σ_{ij} (containing all variances of the components of z_{ij} in its diagonal) is now straightforward. Note that just as equation (17.12) involves the distance between the observation and its expected value scaled inversely by the square of the standard deviation, in our case we use the inverse of the squared covariance matrix, also known as the information matrix (as it denotes the amount of "information" on a variable that is available), which we denote by $\Omega_{ij} = \Sigma_{ij}^{-1}$.

As we are interested in maximizing the joint probability of all measurements $\prod z_{ij}$ over all edge pairings ij following the maximum likelihood estimation framework, it is customary to express the probability density function (PDF) using the log-likelihood. Note that this provides some algebraic convenience in two ways. First, the logarithm is a positive monotonic operation, so the logarithm of any function will not change the points at which it is maximized or minimized. Second, the logarithm of equation (17.12) will result in a linear function in x, which is easier to work with than the cumbersome exponential function. By taking the natural logarithm on both sides of the PDF expression in equation (17.12), the exponential function vanishes and $\log \prod z_{ij}$ becomes $\sum \log z_{ij}$ or $\sum l_{ij}$, where l_{ij} is the log-likelihood distribution for z_{ij}:

$$l_{ij} \propto (z_{ij} - \hat{z}_{ij}(x_i, x_j))^T \Omega_{ij}(z_{ij} - \hat{z}_{ij}(x_i, x_j)). \tag{17.13}$$

Again, the log-likelihood for observation z_{ij} is directly derived from the definition of the normal distribution, but using the information matrix instead of the covariance matrix, it is ridden of the exponential function by taking the logarithm on both sides.

The optimization problem can now be formulated as

$$x^* = \arg\min_x \sum_{<i,j> \in C} e_{ij}^T \Omega_{ij} e_{ij}, \tag{17.14}$$

with $e_{ij}(x_i, x_j) = z_{ij} - \hat{z}_{ij}(x_i, xj)$ being the error between measurement and expected value. Note that the sum actually needs to be minimized as the individual terms are technically the negative log-likelihood.

17.4.2 Numerical Techniques for Graph-Based SLAM

Solving the maximum-likelihood estimate (MLE) problem is nontrivial, especially if the number of constraints provided (i.e., observations that relate one landmark to another) is large. A classical approach is to linearize the problem at the current configuration and reduce it to a problem of the form $Ax = b$. The intuition here is to calculate the impact of small changes in the positions of all nodes on all e_{ij}. After performing this motion, linearization and optimization can be repeated until convergence.

Recently, more powerful numerical methods have been developed. Instead of solving the MLE, one can employ a stochastic gradient descent algorithm. A gradient descent algorithm is an iterative approach to find the optimum of a function by moving along its gradient. Whereas a gradient descent algorithm would calculate the gradient on a fitness landscape from all available constraints, a stochastic gradient descent picks only a subset, and one that is not necessarily random. Intuitive examples are fitting a line to a set of n points but taking only a subset of these points when calculating the next best guess. Because gradient descent works iteratively, the hope is that the algorithm takes a large part of the constraints into account. For solving graph-based SLAM, a stochastic gradient descent algorithm would not take into account all constraints available to the robot but iteratively would work on one constraint after the other. Here, constraints are observations on the mutual pose of nodes i and j. Optimizing these constraints now requires moving both nodes i and j so that the error between where the robot thinks the nodes should be and what it actually sees gets reduced. Because this is a trade-off between multiple and maybe conflicting observations, the result will approximate a maximum-likelihood estimate.

More specifically, with e_{ij} being the error between an observation and what the robot expects to see, based on its previous observation and sensor model, one can distribute the error along the entire trajectory between both landmarks that are involved in the constraint. That is, if the constraint involves landmarks i and j, not only i and j's pose will be updated but all points in-between will be moved a tiny bit.

In graph-based SLAM, edges encode the relative translation and rotation from one node to the other. Thus, to alter a relationship between two nodes, it's necessary to propagate to all nodes in the network. This is because the graph is essentially a chain of nodes whose edges consist of odometry measurements. This chain then becomes a graph whenever observations (using any sensor) introduce additional constraints. Whenever such a "loop-closure" occurs, the resulting error will be distributed over the entire trajectory that connects the two nodes. This is not always necessary. For example, when considering the robot driving a figure-8 pattern, if a loop-closure occurs in one-half of the figure 8, the nodes in the other half are probably not involved.

This can be addressed by constructing a minimum spanning-tree (MST) of the constraint graph. The MST is constructed by doing a depth-first search (DFS) on the constraint graph following odometry constraints. At a loop-closure, (i.e., an edge in the graph that imposes a constraint to a previously seen pose), the DFS backtracks to this node and continues from there to construct the spanning-tree. Updating all poses affected by this new constraint still requires modifying all nodes along the path between the two landmarks that are involved, but inserting additional constraints is greatly simplified. Whenever a robot observes new relationships between any two nodes, only the nodes on the shortest path between the two landmarks on the MST need to be updated.

Take-Home Lessons

• Simultaneous localization and mapping (SLAM) is a key capability for mobile robots to operate autonomously in the world.

• There are robust implementations for environments with strong landmarks—that is, landmarks that can be reliably localized and identified—that use different forms of optimization to find a collection of poses that are most likely, given the available measurements.

• SLAM strongly benefits from additional sensors that can provide additional evidence, in particular beacon-based sensors such as GPS.

• How to deal with environments with dynamical objects—that is, changing maps— remains an open problem.

Exercises

1. In the following, you will develop a basic EKF-based SLAM system with known landmarks:

a) Implement a single-landmark SLAM system. Implement basic odometry in a simulator of your choice as well as a detector to measure the angle and distance to a single landmark. Initialize your first measurement with the mean and variance from your odometry measurement and show how additional measurements can provide a bound on the odometry error using a Kalman filter.

b) Introduce an additional landmark and let the robot return to the first landmark after visiting the second landmark. What can you say about the variance of the second landmark after correcting your variance when reaching the first landmark for the second time?

c) Implement a simulation environment that consists of multiple distinct landmarks. The Ratslife world is a good example (see figure 1.3). Experimentally determine the average variance when localizing against your landmarks. How you do this will depend on the tools already at your disposal. It is OK to cheat, for example, by providing the robot with a list of landmarks close by and simulating a range and bearing measurement. Alternatively, download an open-source SLAM dataset such as the UTIAS Mr.CLAM dataset.

d) Use the tools that you developed above to implement EKF-based SLAM.

2. EKF-based SLAM requires landmarks to be uniquely identifiable. Think about a possible implementation using only corner and wall detectors. How could you make these landmarks appear to be unique, and what is the limitation of this approach?

3. In the following, you will develop a basic graph-based SLAM system:

a) Implement a graph data structure that allows you to maintain pose of the robot and landmarks in the environment. Store translation and rotation from node to node or to the landmark, respectively, on each edge.

b) Implement a DFS algorithm that allows you to compute the shortest path between two nodes on the graph.

c) Use your own simulator or a canned dataset to implement basic graph-based SLAM. Upon loop-closure, update your pose estimate based on the landmark position by averaging between. Use your new estimate to update previous poses along the shortest path back to the previous pose at which the landmark has been observed. Experiment with different policies to update your pose and document your findings.

V APPENDIXES

A Trigonometry

Trigonometry relates angles and lengths of triangles. Figure A.1 shows a right-angled triangle and conventions to label its corners, sides, and angles. In the following, we assume all triangles to have at least one right angle $\left(90 \text{ degrees or } \frac{\pi}{2}\right)$ as all planar triangles can be dissected into two right-angled triangles.

The sum of all angles in any triangle is 180 degrees or 2π, or

$$\alpha + \beta + \gamma = 180°. \tag{A.1}$$

If the triangle is right-angled, the relationship between edges a, b, and c, where c is the edge opposite of the right angle is

$$a^2 + b^2 = c^2. \tag{A.2}$$

The relationship between angles and edge lengths is captured by the trigonometric functions

$$\sin\alpha \quad = \frac{opposite}{hypotenuse} = \frac{a}{c} \tag{A.3}$$

$$\cos\alpha \quad = \frac{adjacent}{hypotenuse} = \frac{b}{c} \tag{A.4}$$

$$\tan\alpha = \frac{opposite}{adjacent} = \frac{\sin\alpha}{\cos\alpha} = \frac{a}{b}. \tag{A.5}$$

Here, the *hypotenuse* is the side of the triangle that is opposite to the right angle. The *adjacent* and *opposite* are relative to a specific angle. For example, in figure A.1, the adjacent of angle α is side b, and the opposite of α is edge a.

Relations between a single angle and the edge lengths are captured by the *law of cosines*:

$$a^2 = b^2 + c^2 - 2bc\cos\alpha. \tag{A.6}$$

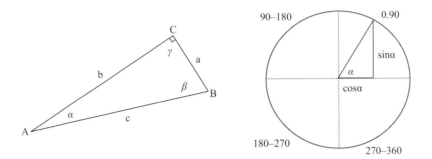

Figure A.1
Left: A right-angled triangle with common notation. Right: Trigonometric relationships on the unit circle and angles corresponding to the four quadrants.

A.1   Inverse Trigonometry

In order to calculate an angle given two edges, one uses inverse functions $\sin^{-1}$, $\cos^{-1}$, and $\tan^{-1}$. (Not to be confused with $\frac{1}{\sin}$.) As functions can, by definition, only map one value to exactly one other value, $\sin^{-1}$ and $\tan^{-1}$ are only defined in the interval $[-90^o; +90^o]$ and $\cos^{-1}$ is defined in the interval $[0^o; 180^o]$. This makes it impossible to calculate angles in the second and third, or the third and fourth quadrant, respectively (figure A.1). In order to overcome this problem, most programming languages implement a function `atan2(opposite,adjacent)`, which evaluates the sign of the numerator and denumerator, provided as two separate parameters.

A.2   Trigonometric Identities

Sine and cosine are periodic, leading to the following identities:

$$\sin\theta = -\sin(-\theta) = -\cos\left(\theta + \frac{\pi}{2}\right) = \cos\left(\theta - \frac{\pi}{2}\right) \tag{A.7}$$

$$\cos\theta = \cos(-\theta) = \sin\left(\theta + \frac{\pi}{2}\right) = -\sin\left(\theta - \frac{\pi}{2}\right). \tag{A.8}$$

The sine or cosine for sums or differences between angles can be calculated using the following identities:

$$\cos(\theta_1 + \theta_2) = \cos(\theta_1)\cos(\theta_2) - \sin(\theta_1)\sin(\theta_2) \tag{A.9}$$

$$\sin(\theta_1 + \theta_2) = \sin(\theta_1)\cos(\theta_2) + \cos(\theta_1)\sin(\theta_2) \tag{A.10}$$

$$\cos(\theta_1 - \theta_2) = \cos(\theta_1)\cos(\theta_2) + \sin(\theta_1)\sin(\theta_2) \tag{A.11}$$

$$\sin(\theta_1 - \theta_2) = \sin(\theta_1)\cos(\theta_2) - \cos(\theta_1)\sin(\theta_2). \tag{A.12}$$

The sum of the squares of sine and cosine for the same angle is one:

$$\cos(\theta)\cos(\theta) + \sin(\theta)\sin(\theta) = 1. \tag{A.13}$$

B Linear Algebra

Linear algebra concerns vector spaces and linear mappings between them. It is central to robotics as it allows describing positions and speeds of the robot within the world as well as moving parts connected to it. Linear algebra is also used in processing image and depth data that are often presented in matrix form.

B.1 Dot Product

The dot product (or scalar product) is the sum of the products of the individual entries of two vectors. Let $hata = (a_1, \ldots, a_n)$ and $\hat{b} = (b_1, \ldots, b_n)$ be two vectors. Then, their dot product $\hat{a} \cdot \hat{b}$ is given by

$$\hat{a} \cdot \hat{b} = \sum_i^n a_i b_i. \tag{B.1}$$

The dot product therefore takes two sequences of numbers and returns a single scalar.

In robotics, the dot product is mostly relevant due to its geometric interpretation

$$\hat{a} \cdot \hat{b} = \|\hat{a}\| \|\hat{b}\| \cos\theta, \tag{B.2}$$

with θ the angle between vectors $\hat{a}$ and $\hat{b}$.

If $\hat{a}$ and $\hat{b}$ are orthogonal, it follows that $\hat{a} \cdot \hat{b} = 0$. If $\hat{a}$ and $\hat{b}$ are parallel, it follows that $\hat{a} \cdot \hat{b} = \|\hat{a}\| \|\hat{b}\|$.

B.2 Cross Product

The cross product $\hat{a} \times \hat{b}$ of two vectors is defined as a vector $\hat{c}$ that is perpendicular to both $\hat{a}$ and $\hat{b}$. Its direction is given by the right-hand rule, and its magnitude is equal to the area of the parallelogram that the vectors span.

Let $\hat{a} = (a_1, a_2, a_3)^T$ and $\hat{b} = (b_1, a_2, a_3)$ be two vectors in R^3. Then, their cross product $\hat{a} \times \hat{b}$ is given by

$$\hat{a} \times \hat{b} = \begin{pmatrix} a_2 b_3 - a_3 b_2 \\ a_3 b_1 - a_1 b_3 \\ a_1 b_2 - a_2 b_1 \end{pmatrix}. \tag{B.3}$$

B.3 Matrix Product

Given an $n \times m$ matrix $\mathbf{A}$ and a $m \times p$ matrix $\mathbf{B}$, the matrix product $\mathbf{AB}$ is defined by

$$(\mathbf{AB})_{ij} = \sum_{k=1}^{m} A_{ik} B_{kj}, \tag{B.4}$$

where the index ij indicates the i-th row and j-th column entry of the resulting $n \times p$ matrix. Each entry therefore consists of the scalar product of the i-th row of $\mathbf{A}$ with the j-th column of $\mathbf{B}$.

Note that for this to work, the right-hand matrix (here $\mathbf{B}$) has to have as many columns as the left-hand matrix (here $\mathbf{A}$) has rows. Therefore, the operation is not commutative (i.e., $\mathbf{AB} \neq \mathbf{BA}$).

For example, multiplying a 3×3 matrix with a 3×1 matrix (a vector) works as follows: Let

$$\mathbf{A} = \begin{pmatrix} a & b & c \\ p & q & r \\ u & v & w \end{pmatrix} \qquad \mathbf{B} = \begin{pmatrix} x \\ y \\ z \end{pmatrix}.$$

Then their matrix product is

$$\mathbf{AB} = \begin{pmatrix} a & b & c \\ p & q & r \\ u & v & w \end{pmatrix} \begin{pmatrix} x \\ y \\ z \end{pmatrix} = \begin{pmatrix} ax + by + cz \\ px + qy + rz \\ ux + vy + wz \end{pmatrix}.$$

B.4 Matrix Inversion

Given a matrix $\mathbf{A}$, finding the inverse $\mathbf{B} = \mathbf{A}^{-1}$ involves solving the system of equations that satisfies

$$\mathbf{AB} = \mathbf{BA} = \mathbf{I}, \tag{B.5}$$

with $\mathbf{I}$ the identity matrix. (The identity matrix is zero everywhere except at its diagonal entries, where its value is one.)

In the particular case of orthonormal matrices in which columns are all orthogonal to each other and of length one, the inverse is equivalent to the transpose. That is,

$$\mathbf{A}^{-1} = \mathbf{A}^T. \tag{B.6}$$

This is important, as rotation matrices are orthonormal.

In case a matrix is not quadratic, we can calculate the pseudo-inverse, which is defined by

$$\mathbf{A}^+ = \mathbf{A}^T (\mathbf{A}\mathbf{A}^T)^{-1} \tag{B.7}$$

and is often used in finding an inverse kinematic solution.

B.5   Principal Component Analysis

Principal component analysis (PCA) breaks n-dimensional data into n vectors so that each data point can be represented by a linear combination of the n vectors. These n vectors have two interesting properties: First, they are ordered by their variance so that the first vector is representative of the data with the highest variation in the data; second, they are orthogonal. These vectors are therefore called *principal components*. Figure B.1 shows an example of two-dimensional data and the two principal components.

This approach has a strong geometrical interpretation: The points along the long axis of the rectangle have higher variance than those along the short axis. Every point in this point cloud can then be reconstructed by a linear combination of the principal component along the long axis and the principal component along the short axis. Finding these vectors is therefore akin to finding the principal axes of the rectangle, regardless of its orientation.

One can show that the principal components are eigenvectors of the data's covariance matrix. For this, we need to compute the mean and variance of data as shown in figure B.1

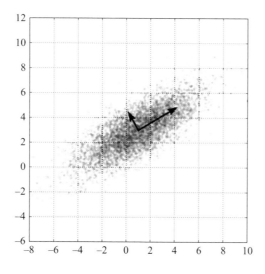

Figure B.1
PCA of a multivariate Gaussian distribution centered at (1,3) with a standard deviation of 3 in roughly the (0.866, 0.5) direction and of 1 in the orthogonal direction. The vectors shown are the eigenvectors of the covariance matrix scaled by the square root of the corresponding eigenvalue and shifted so that their tails are at the mean. *Source*: Nicoguaro CC BY 4.0.

across each dimension, shift the data so that it has zero mean, and then calculate the data's covariance matrix. One can also show that the values of the corresponding eigenvalues are proportional to the importance of each eigenvector.

More formally, given N data samples $\mathbf{x}_i \in \mathbf{R}^n$, we can compute the entries of the $n \times n$ covariance matrix $\mathbf{C}$ as

$$C_{jk} = \frac{1}{N} \sum i = 1^N (x_j^i - \mu_j)(x_k^i - \mu_k), \tag{B.8}$$

with μ_j the mean across the $j - th$ dimension of the data. The eigenvalues λ and eigenvectors $\mathbf{u}$ are given by

$$\mathbf{Cu} = \lambda \mathbf{u} \tag{B.9}$$

and are equivalent to the principal components of the data.

While a typical use of PCA is dimensionality reduction of data (by representing the data only using the n first principal components), PCA is highly relevant in point cloud analysis in robotics—for example, when finding good grasp locations.

C Statistics

C.1 Random Variables and Probability Distributions

Random variables can describe either discrete variables, such as the result from throwing a dice, or continuous variables, such as measuring a distance. In order to learn about the likelihood that a random variable has a certain outcome, we can repeat the experiment many times and record the resulting *random variates*—that is, the actual values of the random variable and the number of times they occurred. For a perfectly cubic dice we will see that the random variable can hold natural numbers from 1 to 6 that have the same likelihood of 1/6.

The function that describes the probability of a random variable to take certain values is called a *probability distribution*. Because the likelihood of all possible random variates in the dice experiment is the same, the dice follows what we call a *uniform distribution*. More accurately, as the outcomes of rolling a dice are discrete numbers, it is actually a discrete uniform distribution. Most random variables are not uniformly distributed, but some variates are more likely than others. For example, when considering a random variable that describes the sum of two simultaneously thrown dice, we can see that the distribution is anything but uniform:

$$
\begin{aligned}
&2 : 1+1 && \rightarrow \tfrac{1}{6}\tfrac{1}{6} \\
&3 : 1+2, 2+1 && \rightarrow 2\tfrac{1}{6}\tfrac{1}{6} \\
&4 : 1+3, 2+2, 3+1 && \rightarrow 3\tfrac{1}{6}\tfrac{1}{6} \\
&5 : 1+4, 2+3, 3+2, 4+1 && \rightarrow 4\tfrac{1}{6}\tfrac{1}{6} \\
&6 : 1+5, 2+4, 3+3, 4+2, 5+1 && \rightarrow 5\tfrac{1}{6}\tfrac{1}{6} \\
&7 : 1+6, 2+5, 3+4, 4+3, 5+2, 6+1 && \rightarrow 6\tfrac{1}{6}\tfrac{1}{6} \\
&8 : 2+6, 3+5, 4+4, 5+3, 6+2 && \rightarrow 5\tfrac{1}{6}\tfrac{1}{6} \\
&9 : 3+6, 4+5, 5+4, 6+3 && \rightarrow 4\tfrac{1}{6}\tfrac{1}{6}
\end{aligned}
\tag{C.1}
$$

$$10 : 4+6, 5+5, 6+4 \quad \rightarrow 3\tfrac{1}{6}\tfrac{1}{6}$$
$$11 : 5+6, 6+5 \qquad\;\; \rightarrow 2\tfrac{1}{6}\tfrac{1}{6}$$
$$12 : 6+6 \qquad\qquad\;\; \rightarrow \tfrac{1}{6}\tfrac{1}{6}$$

As one can see, there are many more possibilities to sum up to a 7 than there are to a 3. While it is possible to store probability distributions such as this one as a lookup table to predict the outcome of an experiment (or that of a measurement), we can also calculate the sum of two random processes analytically (appendix C.3).

C.1.1 The Normal Distribution

One of the most prominent distribution is the Gaussian or normal distribution. The *normal distribution* is characterized by a *mean* and a *variance*. Here, the mean corresponds to the average value of a random variable (or the peak of the distribution), and the variance is a measure of how broadly variates are spread around the mean (or the width of the distribution).

The normal distribution is defined by

$$f(x) = \frac{1}{\sqrt{2\pi\sigma^2}} e^{-\frac{(x-\mu)^2}{2\sigma^2}}, \tag{C.2}$$

where μ is the mean and σ^2 the variance. (σ on its own is known as the standard deviation.) Then, $f(x)$ is the probability for a random variable X to have value x.

The mean is calculated by

$$\mu = \int_{-\infty}^{\infty} x f(x) dx, \tag{C.3}$$

or in other words, each possible value x is weighted by its likelihood and added up.

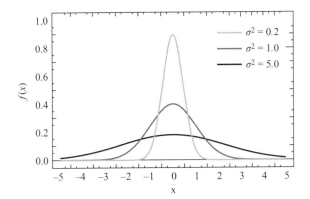

Figure C.1
Normal distribution for different variances and $\mu = 0$.

The variance is calculated by

$$\sigma^2 = \int_{-\infty}^{\infty} (x - \mu)^2 f(x) dx, \tag{C.4}$$

or in other words, we calculate the deviation of each random variable from the mean, square it, and weigh it by its likelihood. Although it is tantalizing to perform this calculation also for the double dice experiment, the resulting value is questionable, as the double dice experiment does not follow a normal distribution. We know this because we actually enumerated all possible outcomes. For other experiments, such as grades in the classes you are taking, we don't know what the real distribution is.

C.1.2 Normal Distribution in Two Dimensions

The normal distribution is not limited to random processes with only one random variable. For example, the X/Y position of a robot in the plane is a random process with two dimensions. In case of a multivariate distribution with k dimensions, the random variable X is a k-dimensional vector of random variables, μ is a k-dimensional vector of means, and σ gets replaced with Σ, a k-by-k dimensional *covariance matrix* (a matrix that carries the variances of each random variable in its diagonal).

C.2 Conditional Probabilities and Bayes' Rule

Let A and B be random events with probabilities $P(A)$ and $P(B)$. We can now say that the probability $P(A \cap B)$ that event A *and* B happen is given by

$$P(A \cap B) = P(A)P(B|A) = P(B)P(A|B). \tag{C.5}$$

Here, $P(B|A)$ is the *conditional probability* that B happens, knowing that event A happens. Likewise, $P(A|B)$ is the probability that event A happens given that B happens.

Bayes' rule relates a conditional probability to its inverse. In other words, if we know the probability of event A happening, given that event B is happening, we can calculate the probability of B occurring given that A is happening. Bayes' rule can be derived from the simple observation that the probability of A and B happening together, ($P(A \cap B)$), is given by $P(A)P(B|A)$ or the probability of A happening and the probability of B happening, given that A happens (equation C.5). From this, deriving Bayes' rule is straightforward:

$$P(A|B) = \frac{P(A)P(B|A)}{P(B)}. \tag{C.6}$$

In other words, if we know the probability that B happens given that A happens, we can calculate that A happens given that B happens.

C.3   Sum of Two Random Processes

Let X and Y be the random variables associated with the numbers shown on two dice (see equation [C.1]), and $Z = X + Y$. With $P(X = x)$, $P(Y = y)$, and $P(Z = z)$ being the probabilities associated with the random variables taking specific values x, y, or z. Given $z = x + y$, the event $Z = z$ is the union of the independent events $X = k$ and $Y = z - k$. We can therefore write

$$P(Z = z) = \sum_{k=-\infty}^{\infty} P(X = k)P(Y = z - k), \tag{C.7}$$

which is the exact definition of a *convolution*, also written as

$$P(Z) = P(X) \star P(Y). \tag{C.8}$$

Numerically calculating the convolution always works and can be done analytically for some probability distributions.

Conveniently, the convolution of two Gaussian distributions is again a Gaussian distribution with a variance that corresponds to the sum of the variances of the individual Gaussians.

C.4   Linear Combinations of Independent Gaussian Random Variables

Let $X_1, X_2, \ldots, X_n$ be n independent random variables with means $\mu_1, \mu_2, \ldots, \mu_n$ and variances $\sigma_1^2, \sigma_2^2, \ldots$, and σ_n^2. Let Y be a random variable that is a linear combination of X_i with weights a_i so that $Y = \sum_{i=1}^{n} a_i X_i$.

As the sum of two Gaussian random variables is again a Gaussian, Y is Gaussian distributed with a mean

$$\mu_Y = \sum_{i=1}^{n} a_i \mu_i \tag{C.9}$$

and a variance

$$\sigma_Y^2 = \sum_{i=1}^{n} a_i^2 \sigma_i^2. \tag{C.10}$$

C.5   Testing Statistical Significance

Robotics is an experimental discipline. This means that algorithms and systems you develop need to be validated by real hardware experiments. Doing an experiment to validate your hypothesis is at the core of the scientific method, and doing it right is a discipline on its own. The key is to show that your results are not simply a result of chance. In practice, this is impossible to show. Instead, it is possible to express the likelihood that your results have not been obtained by chance. This is known as the statistical significance level. How to calculate the statistical significance level depends on the problem you are studying. This section will introduce three common problems in robotics:

1. Testing whether data is indeed distributed according to a specific distribution
2. Testing whether two sets of data are generated from different distributions
3. Testing whether true-false experiments are a sequence of luck or not

C.5.1   Null Hypothesis on Distributions

The null hypothesis is a term from the statistical significance literature and formally captures your main claim. A statistical test can either reject the null hypothesis or fail to reject it. It can never be proved as there will always be a nonzero probability that all your experiments are just a lucky coincidence. The statistical significance level of a null hypothesis is known as the p-value.

The distribution of data is an important class of null hypothesis. Consider, for example, the time it takes to pass a message from one process to another and which follows a log-normal distribution with outliers. We observe three peaks in this histogram. What can we say about message passing times? For example:

- H0: Message passing times follow a Gaussian distribution.
- H0: Message passing times follow a bimodal distribution.
- H0: Message passing times follow a log-normal distribution.

The first null hypothesis implies that message passing takes sometimes a little more time and sometimes a little less, but messages have an average and a variance. The second null hypothesis implies that usually messages take some low average time, but occasionally they are delayed because of the influence of some other process—for example, operating system duties. You can now test each of these hypotheses by calculating the parameters of the distribution to expect and calculate the joint probability that each of your measurements are actually drawn from this distribution. You will find that all of the above hypotheses are almost equally likely. Together, none of your tests will reject your hypothesis. You therefore will need more data.

You can now again calculate parameters for each distribution you suspect. For example, you can calculate the mean and variance of this data and plot the resulting Gaussian distribution. The Gaussian distribution might have a mean slightly offset to the right of the peak. You can also fit the data to a log-normal distribution. You can now calculate the likelihood that the data are actually drawn from either of the two distributions. You will see that the joint probability (the product of all likelihoods) for all data points is actually much higher than that for any Gaussian distribution or any bimodal distribution that you are able to fit.

Formally, this can be done by following Pearson's χ^2-test (read chi-squared test). This test calculates a value that will approximate a χ^2-distribution from all samples and the likelihood of that sample being based on the expected distribution. Plugging the resulting value into the χ^2-distribution leads to the statistical significance level (or p-value).

The value of the test statistic is calculated as follows:

$$\chi^2 = \sum_{i=1}^{n} \frac{(O_i - E_i)^2}{E_i}, \tag{C.11}$$

where

- χ^2 = Pearson's cumulative test statistic, which asymptotically approaches a chi-squared distribution.
- O_i = an observed frequency in the data histogram.
- E_i = an expected (theoretical) frequency, asserted by the null hypothesis (i.e., the distribution you think the data should follow).
- n = the number of samples.

This example also illustrates how statistical tests can be used to determine if you have enough data. If you don't, you will get very poor p-values. In practice, it is up to you to determine what likelihood is significant. Standard significance levels are 10 percent, 5 percent, and 1 percent. If you are unsatisfied with your p-values, you can collect more data and check whether your p-value improves.

C.5.2 Testing Whether Two Distributions Are Independent

Testing whether the data of two experiments are independent is probably the most common statistical test. For example, you might run 10 experiments using algorithm 1 and 10 experiments using algorithm 2. It is up to you to show that the resulting distributions are indeed statistically significantly different. In other words, you need to show that the differences between the algorithm indeed lead to a systematic improvement and that it was not purely luck that one set of experiments turned out "better" than another.

If you have good reasons to believe that your data is normal distributed, you can use a series of simple tests. For example, to test whether two sets of data are distributed with Gaussian distributions that have the same mean, you can use Student's t-test. A generalization of Student's t-test to three or more groups is analysis of variance (ANOVA). These tests have to be done with care as most distributions in robotics are not normal distributed. Examples where Gaussian distributions are commonly assumed are sensor noise on distance measurements such as obtained by infrared or odometry.

If data are not Gaussian distributed, there are other numerical tests for determining the likelihood that two distributions are independent. For example, you could test the message passing time with and without running some computationally expensive image processing routines. You can then test whether the additional computation affects message passing time. If it does, both distributions need to be significantly different. Just using Student's t-test does not work as the distributions are not Gaussian!

Instead, testing whether two sets of data have the same mean needs to be done numerically. A common test is the Mann-Wilcoxon ranked sum test. An implementation of this test is part of most mathematical calculation programs such as Matlab or Mathematica. An extension of the Mann-Wilcoxon ranked sum test for three or more groups is the Kruskal-Wallis one-way analysis of variance test.

C.5.3 Statistical Significance of True-False Tests

There are other experiments that do not lead to distributions but result in simple true-false outcomes. For example, a question one might ask is, "Does the robot correctly understand a spoken command?" This class of experiments is captured by the "lady tasting tea" example. Here, a lady claims to be able to tell the difference in the brewing method of a cup of tea: There is tea prepared by first adding milk and tea prepared by later adding milk. Unfortunately, it is easy to cheat as the likelihood of guessing right is 50 percent. Testing the hypothesis that the lady can indeed differentiate the two brewing methods therefore requires one to conduct a series of experiments to reduce the likelihood of winning by guesswork. In order to do this, one needs to calculate the number of total permutations (or possible outcomes over the entire series of experiments). For example, one could present the lady with eight cups of tea, four brewed one way and four the other. One can now enumerate all possible outcomes of this experiment, ranging from all cups guessed correctly to all cups guessed wrong. There are a total of 70 possible outcomes. Guessing all cups correctly has now a likelihood of 1/70 or 1.4 percent. The likelihood of making a single mistake (16 possible outcomes in this example) is around 23 percent.

C.5.4 Summary

Statistical significance tests allow you to express the likelihood that your experiment is not just the result of chance. There are different tests for different underlying distributions. Therefore, your first task is to convincingly argue what the underlying distribution of your data is. Formally testing how your data are distributed can be achieved using the chi-squared test. The determination of whether two sets of data are coming from two different distributions can then be achieved using Student's t-test (if the distribution is Gaussian) or using the Mann-Wilcoxon ranked sum test if the probability distribution is nonparametric.

D Backpropagation

Simple learning can be achieved by expressing the unknown parameters of a problem in a cost function and then following its gradient to minimize cost. This is straightforward if the cost function is directly differentiable. Calculating the partial derivatives for the error function manually is not straightforward in a multilayer neural network (chapter 10), however, in which a computation graph transforms the input x by a series of multiplications and nonlinear activation functions, which in turn require the chain rule.

Applying the chain rule can be done in two ways: moving forward or backward through the computation graph. Actually doing this by hand for a simple graph shows that going backward is significantly more efficient. Manually deriving the individual partial derivatives also illustrates that many of the computations can actually be recycled. This solution is known as *backpropagation* (Werbos 1990), a technique that has been independently discovered in multiple fields. Because of the technique's relevance beyond training artificial neural networks, it is described in this appendix. The derivation below follows McGonagle et al. (2020). Note that we are using the notation from chapter 10 and section 10.3.2 in particular.

In a first step, we note that the error function is a sum over all input-output pairs:

$$\frac{\partial E(x, y, w)}{\partial w_{i,j}^k} = \frac{1}{2N} \sum_{d=1}^{N} \frac{\partial}{\partial w_{i,j}^k} (\hat{y}_d - y_d)^2 = \frac{1}{2N} \sum_{d=1}^{N} \frac{\partial E_d}{\partial w_{i,j}^k}. \tag{D.1}$$

We will therefore focus on only one input-output pair (x_d, y_d) and differentiate against $w_{i,j}^k$. (The index d has been chosen to avoid confusion with the indices i and j and will be omitted for brevity in the remainder).

The chain rule
The key for understanding the backpropagation algorithm is to apply the chain rule in a correct way. Specifically, if a variable z depends on the variable y, which itself depends on

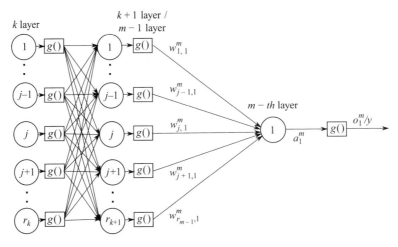

Figure D.1
Last three layers of a neural network with a single output neuron, illustrating dependencies between function values and the output when moving along the computation graph backward.

the variable x, then

$$\frac{dz}{dx} = \frac{dz}{dy}\frac{dy}{dx}. \tag{D.2}$$

With the output layer having index m and a single output (a_1^m), the error is computed by the recursive formula

$$E(x, y, w_{i,j}) = \frac{1}{2}(\hat{y} - y)^2 = \frac{1}{2}(g(a_1^m) - y)^2 = \frac{1}{2}\left(g\left(\sum_{l=0}^{r_{m-1}} w_{l,1}^m o_l^{m-1}\right) - y\right)^2. \tag{D.3}$$

We observe that the variable E depends on the outputs o_l^{m-1} with $l = 0 \dots r_{m-1}$ from the previous layer. Recall that o_l^{m-1} is simply the activation a_l^{m-1} after applying the activation function. Also recall that $w_{i,1}^m$ are weights coming into node 1. The error with respect to $w_{i,j}$ is therefore dependent on all a_j^k for all previous layers. This is also visualized in figure D.1.

The chain rule therefore states that

$$\frac{\partial E}{\partial w_{i,j}^k} = \frac{\partial E}{\partial a_i^k}\frac{\partial a_i^k}{\partial w_{i,j}^k}. \tag{D.4}$$

Error at layer k
The first term is part of a vector called the "error at layer k" that consists of errors at all nodes j in layer k and is denoted by

$$\delta_j^k = \frac{\partial E}{\partial a_j^k}. \tag{D.5}$$

The second term can be computed from the definition of a_j^k above:

$$\frac{\partial a_j^k}{\partial w_{i,j}^k} = \frac{\partial}{\partial w_{i,j}^k} \left(\sum_{l=0}^{r_{k-1}} w_{l,j}^k o_l^{k-1} \right) = o_i^{k-1}, \tag{D.6}$$

which follows from the fact that only the term involving o_i^{k-1} is the one where $l = i$. In case you expect the chain rule to apply further, remember that o_i^{k-1} is actually not dependent on $w_{i,j}^k$, so you are done here.

Thus, the partial derivative of the error function E with respect to weight $w_{i,j}^k$ is

$$\frac{\partial E}{\partial w_{i,j}^k} = \delta_j^k o_i^{k-1}. \tag{D.7}$$

We can see that the error E with respect to each individual weight $w_{i,j}^k$ in a layer k depends on the output of the layers coming before that. This is intuitive, as information propagates through the network. We will now also show that the error term δ_j^k actually depends on the error at layers above k and that it stems from the error $\hat{y} - y$ that we ultimately want to minimize.

D.1 Backward Propagation of Error

To show how the error term δ_i^k relates to the error at the output layer, we will start working backward. Let m be the index of the output layer. We are also only considering a network with one output neuron—that is, $j = 1$. The error at this final layer m is given by

$$E = \frac{1}{2}(\hat{y} - y)^2 = \frac{1}{2}(g(a_1^m) - y)^2. \tag{D.8}$$

Using the chain rule $\frac{\partial E}{\partial w_{i,1}^m} = \frac{\partial E}{\partial a_i^m}\frac{\partial a_i^m}{\partial w_{i,1}^k}$ as before yields

$$\delta_1^m = \frac{\partial E}{\partial a_1^m} = (g(a_1^m) - y)g'(a_1^m) = (\hat{y} - y)g'(a_1^m) \tag{D.9}$$

for the error at layer m and

$$\frac{\partial a_1^m}{\partial w_{i,1}^k} = o_i^{m-1}. \tag{D.10}$$

Together, these two result in

$$\frac{\partial E}{\partial w_{i,1}^m} = (\hat{y} - y)g'(a_1^m)o_i^{m-1}. \tag{D.11}$$

We continue to use the chain rule to work backward along the computation graph. Specifically, the activation a_j^k at node j in layer k, with $1 \leq k < m$, feeds into all nodes $l = 1..r^{k+1}$

of layer $k + 1$. Therefore, the error δ_j^k is calculated as

$$\delta_j^k = \frac{\partial E}{\partial a_j^k} = \sum_{l=1}^{r^{k+1}} \frac{\partial E}{\partial a_l^{k+1}} \frac{\partial a_l^{k+1}}{\partial a_j^k}. \qquad \text{(D.12)}$$

Using $\delta_l^{k+1} = \frac{\partial E}{\partial a_l^{k+1}}$, the above equation simplifies to

$$\delta_j^k = \sum_{l=1}^{r^{k+1}} \delta_l^{k+1} \frac{\partial a_l^{k+1}}{\partial a_j^k}. \qquad \text{(D.13)}$$

Inspecting the computation graph or the definition of a_j^k, we recall that a_l^{k+1} receives the output $g(a_j^k)$ from every node $j = 1..r^k$ in layer k via weight $w_{j,l}^{k+1}$, such that

$$a_l^{k+1} = \sum_{j=1}^{r^k} w_{j,l}^{k+1} g(a_j^k), \qquad \text{(D.14)}$$

allowing us to compute the partial derivative

$$\frac{\partial a_l^{k+1}}{\partial a_j^k} = w_{j,l}^{k+1} g'(a_j^k). \qquad \text{(D.15)}$$

This allows us to provide the error at node j in layer k, also known as the *backpropagation formula*:

$$\delta_j^k = g'(a_j^k) \sum_{l=1}^{r^{k+1}} w_{j,l}^{k+1} \delta_l^{k+1}. \qquad \text{(D.16)}$$

With this last part, we are able to define a recursive definition to calculate the desired error gradient with respect to all weights in the neural network:

$$\frac{\partial E}{\partial w_{i,j}^k} = \delta_j^k o_i^{k-1} = g'(a_j^k) o_i^{k-1} \sum_{l=1}^{r^{k+1}} w_{j,l}^{k+1} \delta_l^{k+1}. \qquad \text{(D.17)}$$

This computation can be executed layer by layer, starting from the output layer and working its way backward. This phase is computationally very similar to the forward phase and allows the reuse of all the activations and outputs that have been previously computed. As an extra goody, the derivative of the sigmoid function $\sigma'(x) = \sigma(x)(1 - \sigma(x))$, resulting in

$$\frac{\partial E}{\partial w_{i,j}^k} = \delta_j^k o_i^{k-1} = g(a_j^k)(1 - g(a_j^k)) o_i^{k-1} \sum_{l=1}^{r^{k+1}} w_{j,l}^{k+1} \delta_l^{k+1} \qquad \text{(D.18)}$$

and from there

$$\frac{\partial E}{\partial w_{i,j}^k} = \delta_j^k o_i^{k-1} = o_j^k (1 - o_j^k) o_i^{k-1} \sum_{l=1}^{r^{k+1}} w_{j,l}^{k+1} \delta_l^{k+1}, \tag{D.19}$$

omitting the need to store a_j^k in addition to o_j^k, reducing the memory requirements of the algorithm by half.

D.2   Backpropagation Algorithm

Training a network now follows these simple steps:

1. Randomly initialize the network's weights.

2. Compute the error for this network for each item in the training set and store the output from each layer (forward propagation).

3. Use the recursive formula for $\frac{\partial E}{\partial w_{i,j}^k}$ to compute the gradient of the error function with respect to each weight using the stored values of the output from forward propagation, and then calculate the average over the entire training set.

4. Repeat steps 2 and 3 for a fixed number of iterations or when the error becomes reasonably small.

Fortunately, calculating the partial derivatives is not very hard in practice since there are tools that automatically calculate the gradient along a computational chain in various programming languages (e.g., autograd, PyTorch, among others). These tools are at the core of modern machine learning frameworks and allow you to construct arbitrary network architectures without worrying about how to actually calculate the gradients. Yet it is difficult to understand how these tools work and what their limitations are without understanding the derivation above.

E   How to Write a Research Paper

The final deliverable of a robotics class often is a write-up of a "research" project, modeled after research done in industry or academia. Roughly, there are three classes of papers:

1. Original research
2. Tutorial
3. Survey

The goal here is to provide guidelines on how to think about your project as a research project and how to report on your results as original research.

E.1   Original Research

Classically, a scientific paper follows the following organization:

1. Abstract
2. Introduction
3. Materials and methods
4. Results
5. Discussion
6. Conclusion

The *abstract* summarizes your paper in a few sentences. Define the problem you want to solve, then explain what method you are employing, what you are doing to assess your work, and what the final outcome will be.

The *introduction* should describe the problem that you are solving and why it is important. A good guideline for writing a good introduction is to use the Heilmeier questions:

1. What are you trying to do? Articulate your objectives using absolutely no jargon.
2. How is it done today, and what are the limits of current practice?
3. What's new in your approach, and why do you think it will be successful?
4. Who cares?
5. If you are successful, what difference will it make?
6. What are the midterm and final "exams" to check for success?

These questions were originally conceived for proposal writing by the head of the Defense Advanced Research Projects Agency (DARPA). There are additional questions to consider as well, but they are left out for the purpose of writing a research paper. For example: "What will it cost?" "How long will it take?" "What are the risks and pay-off?" In the context of scientific research, the question "What are you trying to do?" is best answered in the form of a *hypothesis* (see section E.2).

The *materials and methods* section describes all the tools that you used to solve your problem, as well as your original contribution (e.g., an algorithm that you came up with). This section is hardly ever labeled as such, but it might consist of a series of individual sections describing the robotic platform you are using, the software packages, and flowcharts and other descriptions of how your system works. Make sure you substantiate your design choices using conclusive language or experimental data. Validating these design choices could be your first results.

The *results* section contains data or proofs on how to solve the problem you addressed or why it cannot be solved. It is important that your data is conclusive! You have to address concerns that your results are just a lucky coincidence. You therefore need to run multiple experiments and/or formally prove the workings of your system either using language or math (see also section C.5).

The *discussion* should address limitations of your approach, the conclusiveness of its results, and general concerns that someone who reads your work might have. Put yourself in the role of an external reviewer who seeks to criticize your work. How could you have sabotaged your own experiment? What are the real hurdles that you still need to overcome for your solution to work in practice? Criticizing your own work does not weaken it—it makes it stronger! It becomes clear where the limitations are and where other people can step in with further research.

The *conclusion* should summarize the contribution of your paper. It is a good place to outline potential future work for you and others to do. This future work should not be random stuff that you could possibly think about; rather, it should come out of your discussion and the remaining challenges that you describe there. Another way to think about it is that the "future work" section of your conclusion summarizes your discussion.

It is important not to mix up the different sections of a scientific paper. For example, your results section should exclusively focus on describing your observations and reporting on data (i.e., facts). Don't conjecture here why things came out as they are. You do this either in your hypothesis—the whole reason you conduct experiments in the first place—or in the discussion. Similarly, don't provide additional results in your discussion section.

Try to make the paper as accessible to as many reader styles and attention spans as possible. While this sounds impossible at first, a good way to address this is to think about multiple avenues a reader might take. For example, the reader should get a pretty comprehensive picture of what you do by just reading the abstract, or just reading the introduction, or just reading all the figure captions. (Think about possible avenues because every one you address makes your paper stronger.) It is often possible to provide this experience by adding short sentences that quickly recall the main hypothesis of your work. For example, when describing your robotic platform in the materials section, it does not hurt to introduce the section by stating something like, "In order to show that [the main hypothesis of our work], we selected. . . ." Similarly, you can try reading through your figure captions to see if they provide enough information to follow the paper and understand its main results on their own. It's not a problem to be repetitive in a scientific paper; stressing your one-sentence elevator pitch (or hypothesis) throughout the paper is actually a good thing.

E.2   Hypothesis: Or, What Do We Learn from This Work?

Classically, a hypothesis is a proposed explanation for an observed phenomenon. From this, the hypothesis has emerged as the cornerstone of the scientific method and a very efficient way to organize your thoughts and come up with a one-sentence summary of your work. A proper formulation of your hypothesis should directly lead to the method that you have chosen to test your hypothesis. A good way to think about your hypothesis is that it answers the questions, "What do you want to learn?" or "What do we learn from this work?"

It may be difficult to actually frame your work into a single sentence. What if a single hypothesis seems not to apply? It might be that you are actually trying to accomplish too many things. Can you really describe them all in depth in a six-page document? If yes, maybe some are very minor compared to the others. In that case, they are either supportive of your main idea and can be rolled into this bigger piece of work or they are totally disconnected. If they are disconnected, leave them out for the sake of improving the conciseness of your main message. Finally, you might feel that you don't have a main message but consider all the things you have done to be equally worthy, and despite answering the Heilmeier questions you cannot fill up more than three pages. In this case you might want to pick one of your approaches and dig deeper by comparing it with different methods.

Being able to come up with a one-sentence elevator pitch framed as a hypothesis will actually help you to set the scope of the work that you need to do for a research or class project. How well do you need to implement, design, or describe a certain component of your project? Well, it has to be good enough to follow through with your research objective.

E.3   Survey and Tutorial

The goal of a *survey* is to provide an overview of a body of work—potentially from different communities—and to classify it into different categories using a common language. A survey following such an outline is a possible deliverable for an independent study or a PhD prelim, but it does not lend itself to describe your efforts on a focused research project. Rather, it might result from your involvement in a relatively new area in which you feel important connections between disjointed communities and a common language have not been established.

A different category of survey critically examines concurring methods to solve a particular problem. For example, you might have set out to study manipulation but got stuck in selecting the right sensor suite from the many available options. What sensor is actually best to accomplish a specific task? A survey that answers this question experimentally will follow the same structure as a research paper.

A *tutorial* is closely related to a survey but focuses more on explaining specific technical content (e.g., the workings of a specific class of algorithms or tool), commonly used in a community. A tutorial might be an appropriate way to describe your efforts in a research project, which can serve as an illustration to explain the workings of a specific method you used.

E.4   Writing It Up!

Writing a research report that contains equations, figures, and references requires some tedious bookkeeping. Although technically possible, word processing programs quickly reach their limitations and will lead to frustration. In the scientific community LaTeX has emerged as a quasi-standard for typesetting research documentation. LaTeX is a markup language that strictly divides function and layout. Rather than formatting individual items as bold, italic, and the like, you mark them up as emphasized, section head, and so on, and specify how things look elsewhere. This is usually provided by a template provided by the publisher (or your own). While LaTeX has quite a learning curve compared to other word processing software, it is quickly worth the effort as soon as you need to start worrying about references, figures, or even indices.

Two useful resources are recommended for anyone writing a research report:

- W. Strunk and E. White, *The Elements of Style*, 4th ed. (Longan, 1999).
- T. Oetiker, H. Partl, I. Hyna and E. Schlegl, *The Not So Short Introduction to LaTeX 2_ε*. Available online at https://tobi.oetiker.ch/lshort/lshort.pdf.

F Sample Curricula

This book is designed to cover two full semesters at undergraduate level, CSCI 3302 and CSCI 4302 at the University of Colorado Boulder (CU Boulder), or a single semester "crash course" at graduate level. There are multiple avenues that an instructor can take, each with a unique theme and a varying set of prerequisites for students. Content within the book is deliberately agnostic to a particular robotic platform, programming language, or simulation environment, leaving it to the instructor to choose an appropriate platform.

F.1 An Introduction to Autonomous *Mobile* Robots

This possible one semester curriculum takes the students from the kinematics of a differential-wheel platform to a basic understanding of simultaneous localization and mapping (SLAM). This curriculum is involved and requires a firm background in trigonometry, probability theory, and linear algebra. This might be too ambitious for third-year computer science (CS) students but fares well with aerospace and electrical engineering students, who often have a stronger and more applied mathematical background. This curriculum is therefore also well suited as "advanced class" (e.g., in the fourth year of a CS curriculum).

F.1.1 Overview

The curriculum is motivated by a maze-solving competition that is described in section 1.3. Solving the game can be accomplished using a variety of algorithms ranging from wall following (which requires simple proportional control) to depth-first search on the maze to full SLAM. Here, the rules are designed such that creating a map of the environment leads to a competitive advantage in the long run.

F.1.2   Content

After introducing the field and the curriculum using Chapter 1 (Introduction), another week can be spent on basic concepts from Chapter 2 (Locomotion, Manipulation, and Their Representations), which includes concepts like static and dynamic stability and degrees of freedom. The lab portions of the class can at this time be used to introduce the software and hardware used in the competition. For example, students can experiment with the programming environment of the real robot or set up a simple world in the simulator themselves.

The lecture can then pick up the pace with chapters 2 and 3. Here, the topics of coordinate systems and frames of reference, forward kinematics of a differential-wheel robot, and inverse kinematics of mobile robots are on the critical path, whereas other sections in chapter 3 are optional. It is worth mentioning that the forward kinematics of non-holonomic platforms, and in particular the motivation for considering their treatment in velocity rather than position space, are not straightforward; therefore at least some treatment of arm kinematics is recommended. These concepts can easily be turned into practical experience during the lab session.

The ability to implement point-to-point motions in configuration space thanks to knowledge of inverse kinematics directly lends itself to map representations and path planning as treated in chapter 13. For the purpose of maze solving, simple algorithms like Dijkstra's and A* are sufficient, and sampling-based approaches can be skipped. Implementing a path-planning algorithm both in simulation and on the real robot will provide firsthand experience of uncertainty.

The lecture can then proceed to sensors (chapter 7), which should be used to introduce the concept of uncertainty using concepts like accuracy and precision. These concepts can be formalized using materials in appendix C on statistics and quantified during lab. Here, having students record the histogram of sensor noise distributions is a valuable exercise.

Chapters 8 and 9, which are on vision and feature extraction, respectively, do not need to extend further than needed to understand and implement simple algorithms for detecting the unique features in the maze environment. In practice, these features can usually be detected using basic convolution-based filters from chapter 8 and simple post-processing, for the purpose of introducing the notion of a "feature" but without reviewing more complex image feature detectors. The lab portion of the class should be aimed at identifying markers in the environment and can be scaffolded as much as necessary.

In-depth experimentation with sensors, including vision, serves as a foundation for a more formal treatment of uncertainty in chapter 15 (Uncertainty and Error Propagation). Depending on whether the line fitting example has been treated in chapter 9, it can be used here to demonstrate error propagation from sensor uncertainty and should be simplified otherwise. In the lab, students can actually measure the distribution of robot positions over hundreds of individual trials (this is an exercise that can be done collectively if enough hardware is available) and verify their math using these observations. Alternatively, code to perform these experiments can be provided, giving the students more time to catching up.

The localization problem introduced in chapter 16 is best introduced using Markov localization, from which more advanced concepts such as the particle filter and the Kalman filter can be derived. Performing these experiments in the lab is involved and best done in simulation, which allows neat ways to visualize the probability distributions changing.

The lecture can be concluded with EKF SLAM in chapter 17. Actually implementing EKF SLAM is beyond the scope of an undergraduate robotics class and is achieved only by very few students who go beyond the call of duty. Instead, students should be able to experience the workings of the algorithm in simulation or scaffolded in the experimental platform by the instructor.

The lab portion of the class can be concluded with a competition in which student teams compete against each other. In practice, winning teams differentiate themselves by the most rigorous implementation, often using one of the less complex algorithms (e.g., wall following or simple exploration). Here, it is up to the instructor incentivizing a desired approach.

Depending on the pace of the class in lecture as well as the time that the instructor wishes to reserve for implementation of the final project, lectures can be offset by debates, as described in section F.4.

F.1.3   Implementation Suggestions

An interesting competition environment can be easily re-created using cardboard or LEGO bricks and any miniature, differential-wheel platform that is equipped with a camera to recognize simple markers in the environment (which serve as landmarks for SLAM). The setup can also easily be simulated in a physics-based simulation environment, which allows scaling this curriculum to a large number of participants. The setup implemented at CU Boulder using the e-Puck robot and the open-source Webots simulator is shown in figure F.1.

Variations of the above curriculum can be implemented using a basic Arduino-based platform such as "Sparki." Sparki is equipped with a swiveling ultrasound scanner that can be used to simulate a laser range finder and allows the students to extract simple features such as cones, corners, or gates in the environment and use them for localization. A Bluetooth module allows this robot to be remote controlled; the instructor can move from the Arduino language (C) and computational limitations to a fully-fledged desktop computer.

The class can also be taught using a "Raspberry Pi"-based platform that can be equipped with a webcam, runs Linux, and allows the students to perform basic computer vision using OpenCV and other toolboxes. Here, the Python language and Jupyter Lab provide a low floor to access the programming environment, and a number of educational robots using this architecture have become available recently, some even with graphics processing unit (GPU) support.

Figure F.1
The "Ratslife" maze competition created from LEGO bricks and e-Puck robots (left), and the same environment simulated in Webots (right).

The class can be taught using modified remote controlled (RC) equipped with scanning lasers, stereo cameras, and powerful onboard computation. Competitions among the students can involve decision-making around avoiding obstacles or following a previously unknown course by recognizing landmarks. Descriptions and parts for such vehicles (e.g., the "MIT Racecar") are available online. Here, emphasis will need to change from differential-wheel kinematics to Ackermann kinematics both for odometry and planning.

Finally, a variation of this curriculum can also be taught using drones, such as Parrot drones, which are equipped with a camera as well as a wireless device for executing control algorithms on a desktop computer. In this case, landmarks can be deployed throughout the environment, shifting the focus from kinematics to computer vision.

F.2   An Introduction to Robotic Manipulation

A class on robotic manipulation can be taught at either an introductory or an advanced level, following an introductory course on mobile robots. While teaching autonomous manipulation sets a high bar for linear algebra and vision and feature detection, a manipulation curriculum can also be very practical, shifting the learning experience from the computational into better understanding the role of embodiment.

F.2.1   Overview

A class on robotic manipulation can be inspired by the overview shown in figure 14.1, taking the students from basic arm inverse kinematics to point cloud processing and integrated task and motion planning. By focusing on three-dimensional (3D) perception and inverse kinematics, it is possible to implement the majority of the class in simulation, making the use of a shared hardware resource optional. Alternatively, the class can be taught without any computers and require the students to build their own hardware.

F.2.2   Content

Following the outline of the book, the class can start with mechanisms. Here, the critical role of embodiment should be stressed early on. In chapter 3, the focus is instead on manipulating arms, including the Denavit-Hartenberg scheme and numerical methods for inverse kinematics. In this case, the topics forward kinematics of a differential-wheeled robot and inverse kinematics of mobile robots do not necessarily need to be included. Forward and inverse kinematics can be easily turned into lab sessions using a simple abstraction (Matlab/Mathematica/Python) or simulated robot arm (Webots). If the class uses a more complex or industrial robot arm, an alternative path is to record joint trajectories in a Robot Operating System (ROS) bag and let the students explore this data (e.g., drawing the trajectories recorded from the robot to guess what it has done) before moving on to inverse kinematics.

After introducing forces in chapter 4, theory and practice of grasping can be introduced following the outline in chapter 5.

If the goal of the class is an autonomous solution, the class can then proceed to suitable sensors including basic proprioception, distance sensing, and finally extracting structure from vision. Actuators can be treated as needed, with brushless DC motors and servomotors being standard components of high-performance manipulating systems. If desired, the instructor can also discuss pneumatics and "soft" robotics, which are attractive for manipulating some objects.

With manipulation of a so-called integrated task and motion planning problem, chapter 11 will be an important part of an autonomous manipulation class.

The class can then move on to vision and feature extraction. Topics such as uncertainty and error propagation can be skipped in a class focusing on manipulation. If desired, Bayes' rule can be introduced in the context of "false positives" in object recognition and task execution, allowing the instructor to introduce concepts such as inference in a task planning framework.

F.2.3   Implementation Suggestions

Unless a gripper is provided, designing and modeling a gripper in a robotic simulator can be a worthwhile exercise. Alternatively, students can design their own hardware, 3D-print an end-effector, and try the versatility of their solution by manually actuating their mechanism to solve a set of manipulation challenges such as described in Patel, Segil, and Correll (2016). The sky is the limit here, in particular when "soft" actuators have been introduced and the students are encouraged to compare conventional mechanisms with suction and jamming grippers.

How to teach feature extraction and aspects of mapping will depend on the overarching manipulation goals that are used throughout the class. When focusing on simple bin picking, line recognition and random sample and consensus (RANSAC) can be introduced in the context of identifying the bin and objects therein. In such a scenario, path planning can be substituted by simple inverse kinematics. When focusing on pick-and-place, path planning can be explained by planning around simple obstacles, focusing on rapidly exploring random trees. Labs and experiments can be easily implemented in simulation, initially focusing on perception only and introducing grasp planning only later.

Object recognition and segmentation are good motivation for introducing convolutional neural networks (chapter 10) as well as appropriate open-source tools that the students can use as a black box. Simulators such as Webots also provide object detection and segmentation, allowing the instructor to focus only on the kinematic aspects of autonomy.

Simulation reaches its limitations in tasks that are rich in contacts such as assembly and construction. While a class that is more oriented toward perception might skip this experience, simulation can be complemented by simple experiments in which students create their own hardware. Optimally, a shared resource such an assembly task board is provided that the students can get time on after demonstrating certain basic capabilities in a simulation environment.

F.3   An Introduction to Robotic Systems

A robotic systems class can be offered as an advanced class that allows students to put theoretical concepts to practice, but it can also be taught as a stand-alone class in which advanced concepts are abstracted behind libraries that are presented as "black box."

F.3.1   Overview

A robotic systems curriculum can be motivated by a "grand challenge" task, such as robotic agriculture, robotic construction, or assisted living, all of which require both mobility and a manipulation problem. Although a class project is likely to be limited to a toy example, taking advantage of modern motion-planning frameworks and visualization tools such as ROS/Moveit! (Coleman et al. 2014) makes it easy to put the class into an industry-relevant framework while exposing the students to state-of-the-art platforms in simulation. Possible class projects range from "robot gardening" or "robots building robots," for which setups can easily be created. Examples include real or plastic cherry tomato or strawberry plants and robotic construction kits such as Modular Robotics "Cubelets," which easily snap together to form structures that are robots themselves, adding additional motivation.

F.3.2   Content

The first two weeks of this curriculum can be mostly identical to that described in section F.1.2. If a message passing system such as ROS is used, a good exercise is to record a histogram of message passing times in order to get familiar with the software.

It is now the choice of the instructor whether to focus more on kinematics of arms or differential kinematics for mobile platforms. If the systems class is used in an introductory format, it might be sufficient to introduce basic forward kinematics of robot arms. In an advanced setting, the instructor might instead cover differential kinematics in the force domain.

In case more advanced platforms are available, a depth camera can be mounted above or on the end-effector, then it is possible to introduce topics like vision (chapter 8), feature extraction (chapter 9) and grasping (chapter 5).

F.3.3   Implementation Suggestions

A simple servo-based arm can be mounted on a portable structure that contains a set of fixed (3D) cameras. In order to allow a large number of students to get familiar with the necessary software and hardware, the instructor can provide a virtual machine with a preinstalled Linux environment and simulation tools. In particular, using the Robot Operating System (ROS) allows the recording of "bag" files of sensor values, including entire sequences of joint recordings and RGB-D (color plus depth) video. This way students can work on a large part of the homework and project preparation from a computer lab or from home, maximizing availability of real hardware.

If hardware such as a Kinova arm with integrated Intel RealSense or a Universal Robot arm are available, students can prepare to use the shared resource by working with prerecorded data and a simulation environment. This is not ideal for educating students about grasping, which is not only difficult to simulate but also difficult to understand in a non-experiential setting. The instructor could bridge this gap by letting the students design their own end-effectors using 3D-printing techniques or augmenting simple two-bar linkage grippers with padding (an option that is not explicitly covered in this book). Experimenting with such devices in a remote-controlled setting—as simple as the students manually actuating the gripper—will provide some insights on the challenges of grasping and manipulation. The students can then test their designs with the shared resource and allow the instructor to demonstrate the importance of mechanism and sensing codesign.

F.4   Class Debates

Debates are a good way to decompress at the end of class, create a buffer for students to apply their knowledge by preparing for a final project, and require the students to put the materials they learned into a broader context. Student teams prepare pro and contra arguments for a statement of current technical or societal concern, exercising presentation and research skills. Sample topics may include the following:

• Robots putting humans out of work is a risk that needs to be mitigated.
• Robots should not have the capability to autonomously discharge weapons/drive around in cities (autonomous cars).

• Robots need to be made from components other than links, joints, and gears in order to reach the agility of people.

The students are instructed to make as much use as possible of technical arguments that are grounded in the course materials and in additional literature. For example, students can use the inherent uncertainty of sensors to argue for or against enabling robots to use deadly weapons. Similarly, students relate the importance and impact of current developments in robotics to earlier inventions that led to industrialization when considering the risk of robots putting humans out of work.

Although they may be suspicious at first, students usually receive this format very well. While there is agreement that debates help to prepare students for the engineering profession by improving their presentation skills and prompting them to think about questions posed by society on up-to-date topics, the debates seem to have little effect on changing the students' actual opinions on a topic. For example, in a questionnaire administered after class, only two students responded positively. Students are also undecided about whether the debates helped them to better understand the technical content of the class. Yet students find the debate concept important enough that they prefer it over a more in-depth treatment of the technical content of the class and disagree that debates should be given less time in class. However, students are undecided on whether debates are important enough to merit early inclusion in the curriculum or should be part of every class in engineering.

Concerning the overall format, students find that discussion time was too short when allotting 10 minutes per position and 15 minutes for discussion and rebuttal. Also, students tend to agree that debates are an opportunity to decompress ("relax"), which is desirable when wrapping up the course project.

References

Arkin, R. C. 1989. "Motor schema-based mobile robot navigation." *International Journal of Robotics Research* 8(4): 92–112.

Bay, H., T. Tuytelaars, and L. Van Gool. 2006. "SURF: Speeded up robust features." In *European Conference on Computer Vision*, 404–417. Springer.

Blum, A. L., and R. L. Rivest. 1992. "Training a 3-node neural network is np-complete." *Neural Networks* 5(1): 117–127.

Braitenberg, V. 1986. *Vehicles: Experiments in Synthetic Psychology*. MIT Press.

Brooks, R. A. 1990. "Elephants don't play chess." *Robotics and Autonomous Systems* 6(1–2): 3–15.

Coleman, D., I. Sucan, S. Chitta, and N. Correll. 2014. "Reducing the barrier to entry of complex robotic software: A Moveit! case study." https://doi.org/10.48550/arXiv.1404.3785.

Colledanchise, M., and P. Ögren. 2018. *Behavior Trees in Robotics and AI: An Introduction*. CRC Press.

Correll, N., K. E. Bekris, D. Berenson, O. Brock, A. Causo, K. Hauser, K. Okada, A. Rodriguez, J. M. Romano, and P. R. Wurman. 2016. "Analysis and observations from the first Amazon picking challenge." *IEEE Transactions on Automation Science and Engineering* 15(1): 172–188.

Craig, J. J. 2009. *Introduction to Robotics: Mechanics and Control, 3/E*. Pearson Education India.

Deimel, R., and O. Brock. 2016. "A novel type of compliant and underactuated robotic hand for dexterous grasping." *International Journal of Robotics Research* 35(1–3): 161–185.

Dijkstra, E. W. 1959. "A note on two problems in connexion with graphs." *Numerische Mathematik* 1(1): 269–271.

Dua, D., and C. Graf. 2019. UCI Machine Learning Repository. University of California, School of Information and Computer Science. http://archive.ics.uci.edu/ml.

Duda, R. O., and P. E. Hart. 1972. "Use of the Hough transformation to detect lines and curves in pictures." *Communications of the ACM* 15(1): 11–15.

Ester, M., H.-P. Kriegel, J. Sander, and X. Xu. 1996. "A density-based algorithm for discovering clusters in large spatial databases with noise." In *Kdd-96 Proceedings*, 226–231. Association for the Advancement of Artificial Intelligence (AAAI).

Fikes, R. E., and N. J. Nilsson. 1971. "Strips: A new approach to the application of theorem proving to problem solving." *Artificial Intelligence* 2 (3–4): 189–208.

Floreano, D., and F. Mondada. 1998. "Evolutionary neurocontrollers for autonomous mobile robots." *Neural Networks* 11(7–8): 1461–1478.

Grisetti, G., R. Kummerle, C. Stachniss, and W. Burgard. 2010. "A tutorial on graph-based slam." *IEEE Intelligent Transportation Systems Magazine* 2(4): 31–43.

Harel, D. 1987. "Statecharts: A visual formalism for complex systems." *Science of Computer Programming* 8(3): 231–274.

Hart, P. E., N. J. Nilsson, and B. Raphael. 1968. "A formal basis for the heuristic determination of minimum cost paths." *IEEE Transactions on Systems Science and Cybernetics* 4(2): 100–107.

Hartenberg, R. S., and J. Denavit. 1955. "A kinematic notation for lower pair mechanisms based on matrices." *Journal of Applied Mechanics* 77(2): 215–221.

Henry, P., M. Krainin, E. Herbst, X. Ren, and D. Fox. 2010. "RGB-D mapping: Using depth cameras for dense 3D modeling of indoor environments." In *12th International Symposium on Experimental Robotics*. International Symposium on Experimental Robotics (ISER).

Hughes, A., B. Drury. 2019. *Electric Motors and Drives: Fundamentals, Types and Applications*. Newnes.

Hughes, D., and N. Correll. 2015. "Texture recognition and localization in amorphous robotic skin." *Bioinspiration & Biomimetics* 10(5): 055002.

Katzschmann, R. K., J. DelPreto, R. MacCurdy, and D. Rus. 2018. "Exploration of underwater life with an acoustically controlled soft robotic fish." *Science Robotics* 3(16).

Kavraki, L. E., P. Svestka, J.-C. Latombe, and M. H. Overmars. 1996. "Probabilistic roadmaps for path planning in high-dimensional configuration spaces." *IEEE Transactions on Robotics and Automation* 12(4): 566–580.

Keivan, N., and G. Sibley. 2013. "Realtime simulation-in-the-loop control for agile ground vehicles." In *Conference towards Autonomous Robotic Systems*, 276–287. Springer.

LaValle, S. M. 1998. "Rapidly-exploring random trees a new tool for path planning." Technical Report 98-11. Iowa State University, 1998.

Lowe, D. G. 1999. "Object recognition from local scale-invariant features." In *Proceedings of the Seventh IEEE International Conference on Computer Vision*, vol. 2, 1150–1157. IEEE.

Maulana, E., M. A. Muslim, and V. Hendrayawan. 2015. "Inverse kinematic implementation of four-wheels Mecanum drive mobile robot using stepper motors." In *2025 International Seminar on Intelligent Technology and Its Applications (ISITIA)*, 51–56. IEEE.

Newell, A., J. C. Shaw, and H. A. Simon. 1959. "Report on a general problem solving program." In *IFIP Congress*, 256–264. International Federation for Information Processing (IFIP).

McGonagle, J., G. Shaikouski, C. Williams, A. Hsu, J. Khim, and A. Miller. 2020. "Backpropagation." *Brilliant .org*. https://brilliant.org/wiki/backpropagation.

Nourbakhsh, I., R. Powers, and S. Birchfield. 1995. "Dervish an office-navigating robot." *AI Magazine* 16(2): 53–53.

Otte, M., N. Correll. 2013. "C-forest: Parallel shortest-path planning with super linear speedup." *IEEE Transaction on Robotics* 29(3): 798–806.

Patel, R., R. Cox, and N. Correll. 2018. "Integrated proximity, contact and force sensing using elastomer-embedded commodity proximity sensors." *Autonomous Robots* 42(7): 1443–1458.

Patel, R., J. Segil, and N. Correll. 2016. "Manipulation using the 'Utah' prosthetic hand: The role of stiffness in manipulation. In *Robotic Grasping and Manipulation Challenge*, 107–116. Springer.

Polygerinos, P., N. Correll, S. A. Morin, B. Mosadegh, C. D. Onal, K. Petersen, M. Cianchetti, M. T. Tolley, and R. F. Shepherd. 2017. "Soft robotics: Review of fluid-driven intrinsically soft devices; manufacturing, sensing, control, and applications in human-robot interaction." *Advanced Engineering Materials* 19(12): 1700016.

Pratt, G. A., and M. M. Williamson. 1995. "Series elastic actuators." In *Proceedings 1995 IEEE/RSJ International Conference on Intelligent Robots and Systems. Human Robot Interaction and Cooperative Robots*, vol. 1, 399–406. IEEE.

Rimon, E., and J. Burdick. 2019. *The Mechanics of Robot Grasping*. Cambridge University Press.

Rublee, E., V. Rabaud, K. Konolige, and G. Bradski. 2011. "ORB: An effcient alternative to SIFT or SURF." In *Proceedings of the International Conference on Computer Vision*, 2564–2571. IEEE.

Rusinkiewicz, S., and M. Levoy. 2001. "Efficient variants of the ICP algorithm." In *Third International Conference on 3D Digital Imaging and Modeling (3DIM)*, 145–152. IEEE/RSJ.

Saito, M., H. Chen, K. Okada, M. Inaba, L. Kunze, and M. Beetz. 2011. "Semantic object search in large-scale indoor environments." In *Proceedings of IROS 2012 Workshop on Active Semantic Perception and Object Search in the Real World*. IEEE.

Sibley, G., L. Matthies, and G. Sukhatme. 2010. "Sliding window filter with application to planetary landing." *Journal of Field Robotics* 27(5): 587–608.

Siegwart, R., I. R. Nourbakhsh, and D. Scaramuzza. 2011. *Introduction to Autonomous Mobile Robots*. MIT Press.

Stentz, A. 1994. "Optimal and efficient path planning for partially-known environments." In *Proceedings of the IEEE International Conference on Robotics and Automation*, 3310–3317. IEEE.

Todd, D. J. 1985. *Walking Machines: An Introduction to Legged Robots*. Chapman & Hall.

Van Der Schaft, A. J., and J. M. Schumacher. 2000. *An Introduction to Hybrid Dynamical Systems*, vol. 251. Springer London.

Walter, W. G. 1953. *The Living Brain*. W. W. Norton & Company.

Watson, J., A. Miller, and N. Correll. 2020. "Autonomous industrial assembly using force, torque, and RGB-D sensing." *Advanced Robotics* 34(7–8): 546–559.

Werbos, P. J. 1990. "Backpropagation through time: What it does and how to do it." *Proceedings of the IEEE* 78(10): 1550–1560.

Whelan, T., H. Johannsson, M. Kaess, J. J. Leonard, and J. McDonald. 2013. "Robust real-time visual odometry for dense RGB-D mapping." In *IEEE International Conference on Robotics and Automation (ICRA)*, 5724–5731. IEEE.

Youssefian, S., N. Rahbar, and E. Torres-Jara. 2013. "Contact behavior of soft spherical tactile sensors." *IEEE Sensors Journal* 14(5): 1435–1442.

Zhang, L., B. Curless, and S. M. Seitz. 2002. "Rapid shape acquisition using color structured light and multi-pass dynamic programming." In *Proceedings of the First International Symposium on 3D Data Processing Visualization and Transmission*, 24–36. IEEE.

Index